The Ethics of Tourism Development

Mick Smith and Rosaleen Duffy

Routledge
Taylor & Francis Group

LONDON AND NEW YORK

First published 2003
by Routledge
11 New Fetter Lane, London EC4P 4EE

Simultaneously published in the USA and Canada
by Routledge
29 West 35th Street, New York, NY 10001

Routledge is an imprint of the Taylor and Francis Group

© 2003 Mick Smith and Rosaleen Duffy

Typeset in Times by
Keystroke, Jacaranda Lodge, Wolverhampton
Printed and bound in Great Britain by
MPG Books Ltd, Bodmin

British Library Cataloguing in Publication Data
A catalogue record for this book is available from the British Library

Library of Congress Cataloging in Publication Data
Smith, Mick, 1961–
    The ethics of tourism development / Mick Smith & Rosaleen Duffy.
        p. cm. – (Contemporary geographies of leisure, tourism, and mobility)
    Includes bibliographical references (p.).
    1. Tourism–Social aspects.   2. Tourism–Moral and ethical aspects.
    I. Duffy, Rosaleen.   II. Title.   III. Series: Routledge/contemporary
    geographies of leisure, tourism, and mobility.
    G155.A1S565 2003
    174′.991–dc21   2002155137

ISBN 0–415–26685–8 (hbk)
ISBN 0–415–26686–6 (pbk)

To Alan and Eileen Smith, and
to Margaret and William Duffy

Contents

List of illustrations   ix
Acknowledgements   xi

Introduction   1

1 Ethical values   7

2 The virtues of travel and the virtuous traveller   32

3 The greatest happiness is to travel?   53

4 Rights and codes of practice   73

5 From social justice to an ethics of care   91

6 Authenticity and the ethics of tourism   114

7 Ethics and sustainable tourism   135

8 Conclusion   160

Notes   167
Bibliography   169
Index   190

Illustrations

Plates

2.1 The church of Santo Tomás in Chichecastenango, Guatemala, 1994   37
3.1 Tourists elephant watching in Minneriya National Park, Sri Lanka,
    2001   58
3.2 Another view of tourists in Minneriya National Park   59
3.3 The new five-star Sheraton Hotel in the centre of Addis Ababa,
    Ethiopia, 2002   63
3.4 Whale watching in Tadoussac, Canada, 2000   65
4.1 Dawn at the El Tatio Geyser, Chile, as the tourists disembark, 2000   78
6.1 Idyllic beachscape, Nosy Be, Madagascar, 2001   119
6.2 Sign urging local people to assist in developing the tourism industry,
    Belize City, 2000   121
7.1 Rubbish created by tourism in Placencia being burnt at the rear of
    the main village   136
7.2 Rafael Cal of the Toledo Ecotourism Association, and a TEA
    guesthouse in Medina Bank Village, Belize, 2000   141
7.3 Elephants at Hwange National Park, Zimbabwe, 1996   146
8.1 Independence Day in Cuzco, Peru, 2000   160

Figures

1.1 Spheres of values   10
1.2 Value inter-relations   14
1.3 Relationship between value fields in pre-modern society and in
    modern society   17
1.4 Moral subjectivism regards ethical values as one kind of personal
    preference   25
2.1 'Playing' with social roles and individual values   39
5.1 Two possible distributions of social goods   93
5.2 Rawls suggests that this kind of unequal distribution is acceptable
    because everyone benefits more than in a situation of absolute equality   97

Boxes

1.1   Moral scepticism and the ring of Gyges   20
1.2   Amoralism and moral subjectivism in tourism theory   24
4.1   Tourism development, outdoor pursuits and rights of access   82
4.2   Burma (Myanmar): tourism and human rights   86
5.1   Rawls's original position and real people: the case of the Sa   100
5.2   Tourism and an ethics of care: the ethics of tourism research   108

Acknowledgements

Mick would like to thank the many people who have helped develop and clarify the ideas in this book through discussion and debate. In particular he would like to thank his colleagues in the Unit for Social and Environmental Research (USER) and the division of Sociology at Abertay: Jason Annetts, Gianluigi Giorgioni, Linda Gray, Alex Howson, Alex Law, Wallace McNeish, Andy Panay, Anne Reuss, Peter Romilly and Hazel Work. Between them they have provided the intellectual climate and friendship that still makes academia worthwhile. Thanks also to Eleanor Lothian, with whom I've taught a course on Tourism and Development Ethics for the past four years. I'd also like to mention Phil Mason – sorry I forgot your birthday! Thanks to Sam, Furdie and Horace, and to Joyce Davidson for everything.

Rosaleen would like to thank the many people who were involved in this book. The research was funded by Lancaster University and the Economic and Social Research Council, while I was at Edinburgh University 1997–9 (grant number L320253245) and at Lancaster University 1999–2000 (grant number R000223013). Numerous individuals and organizations provided help and support during the research, not least the tourists who allowed their holidays to be interrupted with interviews. In Toledo District I would like to thank Gregory Ch'oc of the Kekchi Council of Belize, Chet Schmidt of the TEA, Pio Coc of the TMCC and Rafael Cal of the TEA/Medina Bank village. In Belize I was also very grateful for the hospitality shown by Mary Vasquez and Aline Harrison, and thanks to the Marine Research Centre at the University of Belize for providing me with office space and a place to talk over the research. In Zimbabwe I thank Jay Singh, Peter, Melissa and David Makwarimba, and Lauren for their hospitality. When I returned to the UK, my colleagues at Lancaster provided an excellent environment in which to conduct research and they supported me through writing it all up. I remain especially appreciative of the support from Feargal Cochrane while this manuscript was in the final stages of preparation. Finally, I thank Joyce Davidson for her patience over the course of this whole project.

Introduction

Most tourists travel to 'get away from it all', to relax in new surroundings untroubled by the constrictions and irritations that characterize everyday modern life. Leisure is widely regarded as an essential part of contemporary life, and holidays are often presented as necessary to our mental and physical well-being (Krippendorf, 1997; see also Rojek, 1995b). It is, then, both ironic and paradoxical that the tourist industry, which provides so many of us with a means of escape from our mundane existence, should be so dependent upon and epitomize in so many ways this very same modern society. To take a simple example, if tourists are to break from their nine-to-five work routine successfully, to have the 'free time' to 'do their own thing', then those who work in the tourist industry must develop and stick to strictly organized timetables and routines. Planes and transport must be coordinated, waiters be on constant call, food available at short notice, and so on. In other words, the patterns of time management and work that tourists seek relief from are frequently transposed onto the culture of their destination, often in an exaggerated form (Dann, 1996b: 77–79; Markwell, 1997: 138–141).

This transposition is not just a matter of meeting the tourists' expectations – that their holiday should run smoothly, be a home away from home, and so on; it is actually a structural feature of a tourism industry which encapsulates within itself the socio-economic forms and cultural contradictions of late modernity. This is why tourism has proved such an irresistible and important topic for those interested in trying to comprehend the social complexities of contemporary life. MacCannell (1999: xv), in the introduction to the third edition of his now classic text *The Tourist*, states this quite clearly: 'I wanted the book to serve as a new kind of ethnographic report on *modern* society, as a demonstration that ethnography could be directed away from primitive and peasant societies, that it could come home'. Tourists are a 'metaphor of contemporary life' (Bauman, 1997: 93). To study tourism is to study modernity itself, both because, as Urry (1997: 2–3) notes, 'acting as a tourist is one of the defining characteristics of being "modern"' and because tourism is directly responsible for physically exporting the patterns of development associated with modernity worldwide. Tourism, then, can be characterized as an engine and example of patterns of globalization (Duffy, 2002: 127–154; Hoogvelt, 2001; Scholte, 2000). This relationship between globalization and tourism is clear, partly because there is now nowhere, from the Azores to

Antarctica, from Penzance to Papua New Guinea, that has not felt the effects of modernity through tourism development.

While the nature of modernity is itself contested, some arguing that we are now entering a new post-modern situation (Lyotard, 1984), all but the most uncritical advocates of modernization recognize that such developments are deeply ambiguous. They are economically ambiguous in that modernity and development are often heralded as the harbingers of prosperity, yet poverty frequently follows in their wake. They are politically ambiguous in that the freedoms we associate with 'progress', including 'free' time and the freedom to travel, are part and parcel of a society that is ever more ordered and regulated. (Indeed, Rojek (1995b: 2) goes so far as to argue that 'the modernist identification of escape, pleasure and relaxation with leisure was simply another kind of moral regulation'.) They are culturally ambiguous in that modernity envisages itself as breaking with traditional forms of life and yet is dependent upon (re)inventing and (re)visiting such traditions, treating them as resources to provide modern lives with meaning.

The fact that, as Burns (1999) suggests, tourism development continues to be framed by a predominantly Western normative set of values means that these ambiguities become more acute in terms of interactions with the economies, politics and cultures of developing countries, but they are by no means confined to them. As Dicks (2000: 59) argues, heritage is always 'an ambivalent mixture of the authentic and the manufactured', even when that heritage is relatively recent and local, as in the case of mining museums in south Wales. The heritage industry has thus become a site of contested economic, political and cultural values (see, for example, Hale, 2001). As McCrone and co-workers note, Scotland has presented itself in the global tourism market as a brand which reduces Scottishness to tartan shortbread tins and Highland games. This in turn has impacted on attempts to represent Scotland as a modern European nation (Aitchison, 1999; McCrone *et al.*, 1995).

The ambiguities inherent within modernist patterns of economic, political and cultural development also extend to ethics. The question of whether tourism developments are 'good' or bad' is morally charged. 'We now know', says Bauman (1993: 31), 'that we will face forever moral dilemmas without unambiguously good (that is, universally agreed upon, uncontested) solutions, and that we will never be sure where such solutions are to be found; nor even whether it would be good to find them'. If Bauman is right, then there can be no simple answer to whether modernity, development or tourism are 'good' things, partly because late modern society is itself internally divided by sometimes contradictory ethical norms. The contact between modern and other cultures which tourism epitomizes only exacerbates an already morally fraught situation. So we should be clear from the start that a book like this cannot straightforwardly offer solutions to the ethical problems and debates that arise in tourism development. Rather, what we will attempt to do is introduce a number of key ethical discourses – for example, of human rights or virtues – which have been used to frame such debates. Some of these frameworks were explicitly developed to try to address this (modern) situation where universal agreement about what constitutes right and wrong is lacking, and this is the main reason why we concentrate on Western theories of ethics. Utilitarianism is a prime example, since it does not espouse specific, and therefore contestable, moral principles, but instead favours a hedonistic calculus, arguing that our universal desire for happiness alone provides an incontestable basis for moral judgements (see Chapter 3).

Whatever their merits or demerits, each of these discourses frames our ethical problems in a different way, highlighting some aspects while dismissing others as unimportant. The utilitarian concentration on happiness excludes many other aspects of morality, like ideas of duty, integrity, innocence, and so on, that others regard as central to moral debates. Despite claims to the contrary, the ethical frameworks offered by philosophers rarely if ever provide definitive answers to moral problems but are better treated as discursive resources that can help us to articulate and express these problems. From this perspective, a knowledge of ethics is not like a knowledge of mathematics, it will not allow us to 'solve' complex social equations simply, but it might help us interpret and communicate to others what it is that we think is right or wrong about a certain situation and why. One of our key aims is therefore to introduce these ethical discourses in such a way that their relevance to debates about tourism development becomes apparent. But this book is not simply, or even primarily, an exercise in what is sometimes termed 'applied philosophy' – that is, in taking abstract philosophical theories and then applying them to a given issue, in this case tourism. Rather, we try to provide a sociologically and politically grounded account of these ethical theories and show how they too might be implicated in the (re)production of particular social circumstances and moral problems.

So far as we are aware, this text is the first to deal explicitly and in its entirety with the ethics of tourism development. However, recent years have certainly seen a renewed interest in moral matters in many academic disciplines and in many practical contexts. This 'turn to ethics' (Garber *et al.*, 2000) has been exemplified in the number of new journals with an ethical focus now on the shelves of university libraries. David Smith (2000: 7), for example, has recently argued that the '(re)discovery of moral issues by geographers may be linked to a broader normative turn in social theory', and to increased interest in the relations between ethics and economics, urban planning and anthropology. These developments, together with relatively new fields like environmental ethics, are of direct relevance to the study of tourism.

We hope that this book will demonstrate that it is impossible to separate ethics from the question of development in general and tourism development in particular. Development is much more than an economic process: it is driven by particular ideas, values and norms, all of which can be brought into question. Such differences in values are most obvious in cases where the 'developed' North meets the supposedly under-developed South, and for this reason many, but by no means all, of our examples are taken from what are sometimes referred to as 'Third World' countries such as Zimbabwe and Belize. Here again, though, we need to recognize that even the terminology used to describe the development process carries within it implicit assumptions and values. Thus Escobar (1995) argues that the terms 'First World' and 'Third World' are themselves categories constructed by the developed North to allow it to intervene in the South to 'correct' the latter's problems.

Such interventions characteristically take the form of neoliberal modernization and development policies that are supposed to be universally applicable (see Chandler, 2002). The central core of these strategies is an emphasis on economic diversification, particularly a commitment to non-traditional exports such as tourism (Brohman, 1996: 48–52). Southern countries are perceived to have a comparative advantage in terms of selling sun, sea, sand and sex, thereby fitting with Northern tastes for what seem like environmentally unspoiled and culturally exotic destinations. This neoliberal approach has also been favoured by the international lending agencies such as the World Bank and the IMF, and by bilateral donors which have made loans available in return for reforms that favour market-oriented growth (Biersteker, 1995; World Bank, 1994). These institutions are in effect implementing theories of modernization that emerged after 1945, perhaps most closely associated with Samuel P. Huntington and Walt Rostow (Huntington, 1968; Rostow, 1991). The spread of the so-called 'Washington consensus' around the world means that the switch from command to market economies also involves a package of reforms that require countries to begin democratization and to establish systems to ensure good governance. In this way the developing world is urged to mimic the 'First World', so that it can 'catch up' through adoption of the same kinds of economic and political management techniques. The usually unspoken ethical and political assumption here is that this global dominance of liberal ideas and capitalist economics constitutes the apex of human history (Fukuyama, 1992). More recent debates about globalization have only served to accentuate this theme of modernization. While some proponents of globalization have pointed to its ability to encourage economic modernization in the developing world, critics have characterized globalization as a homogenizing force (see Hoogvelt, 2001; Scholte, 2000). Still, it is clear that the effects of globalization are extremely uneven and induce changes that benefit some political, economic and social groups while disadvantaging others. The unevenness of tourism developments, as part of broader processes of globalization, provides an ideal vantage point from which to analyze the ethical dimensions of globalization itself.

Governments in the South facing financial problems and an end to secure markets for their goods in former colonial powers have in turn recognized that tourism provides a potential answer to their problems (Lynn, 1992: 371–373). Most, regardless of their political ideology, promote tourism as a means of generating employment and a healthy balance of payments, and as a source of foreign exchange (Hall, 1994: 112–120; Harrison, 1992b: 8–11). Many governments are keen to justify public-sector tourism development through the use of a pro-business rhetoric (Clancy, 1999; Matthews and Richter, 1991: 120–123). In this way, national tourism policies tend to focus on economic growth, and it is regarded as synonymous with Westernization and modernization. The spread of a neoliberal market philosophy goes hand in glove with the increasing hegemony of specifically Western cultural and political values.

This prioritizing of economic considerations is thus not part of a politically neutral approach to development but is a reflection of modernity's own situation. We hope to raise awareness of this and to refocus debates around tourism development on the aesthetic, cultural, environmental and, most of all, ethical values of landscapes and cultures which the neoliberal development paradigm largely fails to take into account. Tourism development too is ambiguous. The benefits it brings can come at a cost in terms of the erosion of traditional cultures, environmental degradation, exploitation and social fragmentation. For example, tourism is often criticized for leading local people to attempt to mimic the consumption patterns of the tourists. This is commonly called the demonstration effect, where the values of materialism and hedonism, imported by tourists, creates a demand for Western lifestyles and attitudes among hosts. Harrison suggests that the demonstration effect is the cultural equivalent to the spread of market relationships and commoditization, and hence of neoliberalism (Harrison, 1992c: 29–31; Lynn, 1992: 371–377; Pattullo, 1996: 84–90). It is also interesting to note that the call for more environmentally aware and culturally sensitive forms of tourism has itself been informed by critiques of the dominant development strategy from within dependency theory, world systems theory and post-development theories (see Hoogvelt, 2001; Scholte 2000; Wallerstein, 1979).

In an age of increasing globalization, *The Ethics of Tourism Development* seeks to elucidate the inter-relations between modernity, ethics and tourism, and account for the manner in which tourism, often all too uncritically, exports the values and presuppositions associated with its own cultural origins. We begin in Chapter 1 by making a case for taking ethical values seriously over and against the doubts generated by, among other things, the predominance of economic evaluations in modern societies. Ethical values, unlike economic values, often seem intangible, difficult to pin down and impossible to measure. But, we argue, these values are still vital aspects of our everyday lives and of cultures in general. In Chapter 2 we turn to the question of moral relativism. How is it possible to make ethical judgements about right and wrong in a world with widespread differences in moral values? This leads into a discussion of the importance of what might be termed 'moral character': the virtues that we might expect individuals to cultivate and exhibit in their dealings with others. Tourism has often been promoted as character-building in terms of developing virtues like independence, tolerance, and so on. Successive chapters then examine the two dominant modern ethical frameworks, namely utilitarianism and human rights, which claim to provide modernity's own common currency for addressing ethical issues. Can such discourses provide a way of conceptualizing the problems associated with issues like sex tourism or travel to countries with oppressive regimes? In Chapter 5 we consider the ethical and political issues arising in many areas of tourism development in terms of questions of social justice, discourse ethics and what has come to be known as an ethics of care. Here we examine the profound cultural and environmental impacts on tourist destinations, the reciprocity (or lack of reciprocity) in 'host'–'guest' relations and the (un)fair distribution of benefits and revenues. Chapter 6 turns to the moral implications of issues like 'staged authenticity'. The book concludes with a detailed investigation of the potential and pitfalls of ecotourism, sustainable tourism and 'community-based tourism' as examples of what has been referred to as 'ethical tourism'. The final two chapters attempt to bring together much of the theoretical material discussed earlier through detailed discussions of the moral complexities revealed during the course of primary research into tourism development in Belize and Zimbabwe.

1   Ethical values

As we enter the twenty-first century, even the highest and least hospitable spot in the world, the top of Mount Everest, has become something of a tourist destination. Interested parties can pay up to US$65,000 to one of several competing commercial companies in an attempt to get their name in the roll call of those who have reached its summit. In the spring of 1996 no fewer than thirty expeditions jostled for position at the base of the mountain, and on one day alone some forty climbers actually managed to reach the top. Yet as Joe Simpson argues in *Dark Shadows Falling*, this commercialization of the ultimate mountaineering experience has not removed its dangers and not come without a cost. By 1997 150 had died in the attempt, including fifty sherpas. The upper slopes of Everest have become littered with frozen bodies, old oxygen tanks and the unsightly clutter of numerous expeditions. There are even accounts of how, in their desperation to reach the summit, climbers have simply walked past still-dying members of previous expeditions, making no attempt to help them down or even just to stay with them in their last moments. One such climber was quoted as saying, 'above eight thousand metres is not a place where people can afford morality' (Shigekawa in Simpson, 1997: 48). Simpson disagrees:

> I find it unforgivable that climbers can treat their fellow mountaineers with such callous disregard. It has nothing to do with whether or not rescue appears to be possible but everything to do with being humane, caring individuals who can see the passing of a life for what it is, and not simply an inconvenient obstacle to realising egoistic ambitions. . . . If climbers 'cannot afford morality', and ethical behaviour becomes too expensive, then has the sport become prostituted?
>
> (1997: 198, 200)

These are extreme examples from an extreme environment, but they raise pertinent points for any account of tourism development. Is tourism all about the egoistic satisfaction of those paying for the privilege or should ethics play a part? What does it mean to say that a certain way of behaving, or a particular kind of tourism development, is wrong? Can the tourism industry 'afford' morality?

There are no easy answers to such questions. Yet despite the frequency of our everyday use of ethical terms to describe certain states of affairs as 'right' or 'wrong', 'good' or 'evil', 'just' or 'unjust', and so on, such evaluations often seem to carry little weight with key decision makers. This ethical 'deficit' often seems to be a feature of many areas of modern society such as public policy, scientific research or business developments. Some people even doubt the applicability of ethics to such areas at all. They claim that public policy is purely a matter of political pragmatism, that scientific research is objective and value free and should remain unfettered by ethical considerations, and that the role of business is simply to make bigger profits. Although it is unusual to hear these views expressed quite so strongly, there is certainly a spectrum of opinion about the importance and relevance of ethical values. So before we can even begin to discuss theories of 'virtue', 'rights', 'justice', and so on, and try to understand their implications for tourism development, we need to address those who are sceptical about ethics' importance, who doubt its claims or deny its relevance. Perhaps the easiest way to do this is by trying to compare and contrast ethical with other kinds of values.

Ethical, economic and aesthetic values

The playwright Oscar Wilde famously defined a cynic as someone 'who knows the price of everything and the value of nothing'. As usual, Wilde's (cynical) witticism is much more than mere word-play; its humour stems from its accuracy as a form of social commentary. It often seems that the modern world is dominated by economic concerns to the exclusion of all other values. When it comes to deciding a course of action, 'money talks', and profit is frequently the 'bottom line'. Financial matters dominate every aspect of our lives; we must work to earn money and we need money to afford the time off work to 'play'. Given its central role in our everyday lives, it is easy to forget that 'money isn't everything', that there have been numerous societies that have existed entirely without a cash economy, and that, as Wilde suggests, there may be values in our own society that simply cannot be accounted for monetarily.

Of course, the fact that we can account for economic values, that we can price things, is one of the reasons that economic values have assumed their current importance. Although we recognize that in one sense money is only symbolic – its value depends on a kind of social agreement to treat it as something of value (if you don't believe this, try spending Albanian leks or Algerian dinars in an Edinburgh supermarket) – money does serve to make economic values *tangible*. In Simmel's (1990: 172) words, 'money is entirely a sociological phenomenon, a form of human interaction'. Every day we grasp the paper, coins and plastic that represent our financial clout in our hands, we exchange quantifiable amounts of cash for items that we can eat, sleep on or drive. Through its function as a means of exchange of diverse forms of goods, from food to furniture, and cleaning services to cars, money becomes that currency which connects together every aspect of the production and consumption of goods in contemporary society.

Despite – indeed, because of – the fact that money provides a very abstract measure of something's economic value, it also seems to provide us with a very concrete way of expressing that thing's value as *hard* currency, as being worth so many dollars, yen or pounds.

Given money's flexibility and its apparently unlimited ability to encapsulate the value of so many different things, it is understandable that modern society tends to put a price on everything, to turn everything into the form of a commodity that, at least potentially, might be bought or sold. By comparison, the kind of values to which Wilde refers, the values that he claims are excluded from or beyond price, seem extremely intangible. They are not easily quantifiable – if indeed they are quantifiable at all. How do we go about measuring the amount of love we feel for someone or how much we care for our parents or our pets? How can we quantify the beauty of a sunset? These kind of ethical and aesthetic values seem to be quite different from economic values; they have little to do with exchanging (consumer) goods but concern themselves with 'goods' of a different kind, such as with living an 'ethically' good life or painting an 'aesthetically' good painting.

In other words, we seem to have at least three different ways of valuing things, and each form of evaluation uses quite different criteria. Roughly speaking, economics (from the ancient Greek word *oikos*, meaning pertaining to the house-hold) concerns itself with valuing those material things we produce and consume in everyday life. Aesthetics (again from an ancient Greek word, *aestheta*, meaning pertaining to the senses or matters of taste) concerns itself with appreciating and passing judgement about a thing's beauty. Ethics (from the Greek *ethika*) concerns itself with evaluating the moral worth of a thing, an action or a person. The point that Wilde's remark makes so forcibly is that merely knowing something's economic value tells us nothing about, and may even detract from knowledge about, what he regards as these other, far more fundamental forms of evaluation.

We hope that the examples of tourism, many in the developing world (or the 'South'), that we look at in this book will assist us in examining the complex inter-relationship between ethical, aesthetic and economic values. Later chapters will examine debates about how economics and material values affect tourism policies from the perspective of the various interest groups involved. We will return to the role that aesthetics plays in determining the ways that tourists view the landscapes, cultures and histories that they gaze upon. Most importantly, we will analyze the moral values that are a central focus of some of the locally organized and managed sustainable tourism projects in the developing world.

We could, of course, differentiate many more than these three kinds of value, and these different forms of value are not always entirely separate or separable. For example, there are 'religious values' which in some cases can overlap with and even underpin certain ethical values. It would be difficult, if not impossible, to understand the moral values espoused within, say, a catholic community or those embodied in the classical Hindu caste system without some knowledge of their respective religious beliefs. Knowledge and respect for such values is a vital part of any responsible tourism. For example, Gehrels (1997) recounts the case of tourists' negative impacts on the nomadic Himba group of north-western Namibia.

'In the past many tourists carelessly or unwittingly violated the Himba's sacred sites, crossing "spirit lines" and driving across old Himba encampments for example.' Such problems created severe tensions that have only been averted as the Himba have regained some control over tourism on their lands. We might also recognize what could be termed 'epistemic' values, those concerned with the quality and value of the knowledge we have. We do, after all, usually value the truth quite highly. But for our present purposes we will avoid these complications and suggest that we might picture economics, ethics and aesthetics in terms of three distinct but overlapping value spheres (see Figure 1.1).

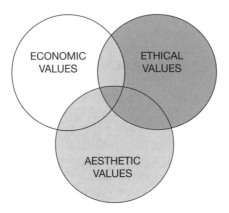

Figure 1.1 Spheres of values

This model obviously oversimplifies matters, but it is useful to begin by teasing these different kinds of values apart, not least because this book primarily concerns itself with only one of these value spheres, that of ethics. As we shall see, this does not mean that we will treat ethics in isolation from economic, aesthetic or indeed any other aspects of society. To do so would in any case be impossible in the context of tourism development since economic and aesthetic values interact with ethical values in very complex ways. But it is precisely for this reason that we need to establish an understanding about the scope and nature of the differences between ethical, aesthetic and economic values.

Perhaps we can best illustrate these issues in relation to a particular example of a specific case of tourism development. In the 1950s many of those regions of the Spanish Mediterranean coast now so familiar as package holiday resorts were relatively undeveloped. Life in the more remote villages that dotted the coast went on much as it had for generations. The travel writer Norman Lewis (1984), who returned to one such village he called Farol over a period of several years, described how traditional village life revolved around seasonal fishing for sardines and tunny, the calendar of church festivals, and respect for the authority of the village *alcalde*, or mayor. But the arrival of tourists, primarily from Northern Europe, began a process of development that was to completely change such villages in a matter of a few short years.

It is worth examining Lewis's account of the changes he witnessed taking place, because they exemplify what Young (1983) refers to as the 'touristization' of the traditional Mediterranean fishing village and are typical of many of the value issues raised by tourism development in general. Lewis describes how a local 'entrepreneur' whom he calls Muga first bought up the decaying local mansions, one of which he converted into a hotel, covering it with imported tiles and festooning it with fairy lights. He then rapidly began preparing the way for tourism, building an incongruous Moorish-style café, removing all the messy apparel of the fishermen's daily activities from the waterfront so as to sanitize the tourists' view, and building a new road made of concrete and covered with 'sackloads of imitation marble chippings of many clashing colours' (Lewis, 1984: 153). Thus Farol, Lewis says,

> began its slow loss of identity, Muga went from strength to strength, busied with his plans for the coming of the tourists, determined to create for them here a Spanish dreamland, a gimcrack Carmen setting in which the realities of poverty and work were tolerable so long as they remained picturesque.
>
> (1984: 152–153)

Having first employed villagers to work as cleaners and cooks in his hotel, Muga next tried to persuade the fishermen to work for him, running tourist boat-trips. The fishermen, Muga argued, usually made only thirty-five pesetas for a day's work.

> 'For this', Muga said, 'you put in eight hours' hard work, two hours to put the nets down, two hours to take them up, and four to get the fish out of the nets and box them up. I don't want to talk about all of the time that goes into mending the nets, especially when the dolphins have been at them. At Palamos they take tourists for a two- or three-hour boat trip, and they're paid 1,000 pesetas. Quite a difference isn't there? . . . This is your chance,' he said. 'You can charge 1,000 pesetas for taking a party to one of the beaches. Why bother about sardines? Why bother about tunny?' The fishermen's expressions made it clear they were horrified, that Muga's proposal seemed to them immoral, almost indecent. . . . What Muga now suggested, they complained, was an affront not only to them, but to the sea. They were *gente honrada* [honorable people], not tourist touts or pimps.
>
> (Lewis, 1984: 155)

Although Lewis's account may be somewhat embellished for dramatic effect, innumerable academic studies from many parts of the world have highlighted exactly the same kind of issues in relation to tourism development in traditional societies (Mansperger, 1995; see also Abbink 2000; Butler and Hinch, 1996; Price, 1996). The changes that occurred in Farol raise matters that relate directly to differences in aesthetic, ethical and economic values between those proposing the tourism development and those wishing to retain at least something of their more traditional ways of life. Lewis cites numerous examples of differences in aesthetic values between Muga, who is trying to make the village conform to tourists' expectations, and the villagers themselves. These include the cafés' and hotels' design, which confounds local tastes by introducing new and quite alien architectural elements; the 'prettification' of the seafront; and the new road covered in multicolour chippings that Lewis's neighbour says makes him 'have the feeling of having eaten something I cannot digest' (Lewis, 1984: 153).

The economic values Muga espouses could not be expressed more clearly. There are tangible economic advantages for the fishermen and their families to abandon traditional working practices in favour of work servicing the tourist industry. A thousand pesetas a day is clearly far more than the thirty-five they previously earned. The development also has other material advantages, including massive improvements in transport infrastructure and bringing piped water and telecommunications to the village for the first time.

Yet these economic advantages also depend upon the villagers accepting drastic changes to their way of life, their culture and even their personal identities. In making such direct financial comparisons Muga explicitly turns the fishermen's time and labour into a 'commodity', something to be bought or sold.[1] The fishermen are, in effect, being asked to decide whether to become (relatively) well-paid wage-labourers within a global economic system or to retain their relatively autonomous existence and current social status as much poorer participants in a traditional subsistence economy. We use the word 'decide' here advisedly, since, because of the financial and political clout of developers like Muga, there may not actually be much choice involved. Developers can (and often do) force changes through whether or not the majority of the local population actually support them (as Chapter 7 clearly shows; see also Burns, 1999; Duffy, 2002: 98–126; ESTAP, 2000).

Even this very simple example raises a series of complex ethical questions that are inextricably interwoven with the social and political fabric of local cultures and, where tourism development is concerned, global systems. What Muga proposes as a matter of simple and straightforward economics actually has wide-ranging social repercussions that clearly conflict with the fishermen's traditional cultural practices and deeply held beliefs. They are horrified that anyone should suggest that they should sell themselves (their labour) or their heritage in this 'immoral' way, like 'touts or pimps'. They employ moral language to show that they regard the suggestion as an offence to their dignity and even an affront to the sea itself. The idea of affronting the sea is interesting because it suggests that what the fishermen deem to be wrong here is not just the commodification of their own labour but the treating of the sea as a commodity, as a resource that can be bought and then sold to tourists. This point is emphasized from an anthropological point of view by Greenwood (1989: 174), who argues that the 'commoditization of local culture in the tourism industry is . . . fundamentally destructive and . . . the sale of "culture by the pound", as it were, needs to be examined by everyone involved in tourism' (see also Philip and Mercer, 1999; Watson and Kopachevsky, 1998). Interestingly, this process of commodification eventually affects everyone concerned; even the tourists themselves eventually come to be seen by locals as nothing more than a resource (Pi-Sunyer, 1989: 197).

In expressing these moral qualms the villagers voice what they regard as ethical limits on the way in which economic values should be employed; they are (indirectly) opposing the spread of the market into areas which their traditional culture leads them to believe should remain inviolable. The fact that the villagers are uncomfortable with, and try to resist, the economization of certain aspects of their lives seems to support our characterization of ethics as an autonomous field of values (Figure 1.1), implying that ethical values are different from and not reducible to economic values. Similarly, the fact that the villagers retain their aesthetic opinions about Muga's 'ugly' architectural 'enhancements', despite the fact that they are likely to be economically beneficial, also seems to point to an autonomous field of aesthetic valuation.

However, it is important to qualify this argument in several ways. We should recognize that economic, ethical and aesthetic values are not entirely separate: they do interact, and the boundaries between them are not hard and fast. For example, it is easy to see how Muga might employ moral arguments himself to try to convince the fishermen of the advantages of his tourist development, perhaps by pointing out their ethical obligations to support their families and provide for them as best they can. In this case Muga would be making a moral argument in favour of the fishermen jettisoning their previous ethical objections so that they and their families might benefit economically. This kind of interaction also leads to the blurring of boundaries between the value fields. For example, someone strongly opposed to Muga's proposals might regard the money he offers to take tourists on boat-trips as a 'bribe' rather than a straightforward economic transaction. The term 'bribe' has obvious moral connotations suggesting that Muga's money is (ethically) tainted. Similar examples might show how aesthetics too interacts with and is influenced by ethics and economics. Thus we have to recognize that these value-fields are only ever *relatively* autonomous; they are not completely self-contained and their degree of autonomy varies in different circumstances.[2] We could represent this pictorially as in Figure 1.2, which indicates that economic, ethical and aesthetic values not only overlap, but are inter-related and impact upon one another.

As Figure 1.2 illustrates, thinking of these value-fields as only 'relatively' autonomous might also help to counter certain philosophical tendencies to try to define ethics as though it were a timeless and unchanging realm with fixed contents.[3] As the Introduction has already suggested, if we draw our definition of ethics too tightly and impose rigid boundaries on what is, or is not, supposed to be a matter for ethics, we inevitably end up removing morality from the social contexts that give it meaning. There is no way of predefining what can and cannot be treated as an ethical issue and no way of understanding why something is counted as an ethical issue without looking at its particular social context.

For example, those of us who already inhabit a society where wage-labour is prevalent might think the fishermen's reaction to Muga's offer of employment very odd indeed. Surely there can't be an ethical issue here? Many of the people

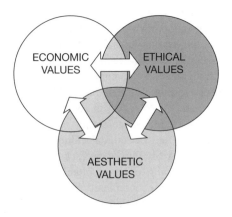

Figure 1.2 Value inter-relations

reading this book will, after all, be seeking employment as wage-labourers in the tourist industry, and this prospect will not have troubled their conscience or offended their dignity. But this is to miss the point. We live in a culture where wage-labour has become the norm, where indeed we are actually under moral as well as economic pressures to obtain 'gainful' employment. (Sociologists like Max Weber (2001) have, of course, argued that an acceptance of this 'work ethic' was a vital ingredient in the development of capitalist society.) The fishermen, however, draw upon an entirely different cultural context. They are not used to selling their labour in this way or to being at the beck and call of tourists or tour operators. For generations they have run their lives around the cycles of nature and tradition: the arrival of the sardine shoals, the effects of the weather, the festivals celebrating a good catch or the tales in the tavern of a disappointing day's fishing.[4] To abandon all this for money is, to them, comparable to selling their souls. It threatens to overthrow their relations to tradition, to overturn a community where fishing plays a (the) key role and to undermine their sense of self-identity since they define themselves and are defined by others by their position as fishermen. Refusing Muga's money is therefore regarded as a matter of cultural and individual 'integrity', in both the moral and the non-moral sense of the word.

In other words, the ethical issues that arise in many, if not most, cases of tourism development are closely connected to the socio-economic effects that development has on the 'host' community. The greater the disparity between the host community's social organization and that associated with the tourist development, the more scope there is for the introduction of novel practices and behaviours to lead to value conflicts. Of course, it is not inevitable that value conflicts will follow in the wake of tourism development. It would be naïve to view tourism as always detrimental. Some communities may welcome change, and, as we shall see, much can be done to defuse and ameliorate potential problems. Abbott-Cone (1995), Brown (1992), Pearce (1995) and Tosun (2002) all demonstrate that local cultures respond in a variety of ways to the arrival of tourism developments. In each community, state or region, host cultures respond to tourism in multiple ways. In the same community some interest groups are able to engage with tourism in a proactive way that benefits them, while other groups react more passively, deriving few benefits from, or being disadvantaged by, the new developments. However, it is important to recognize that ethical disputes are much more than simply differences of opinion about what is right and wrong (see below); they are rooted in different 'forms of life' and indicative of different social expectations and aspirations.[5]

Lewis's account makes it clear that tourism brings profound socio-economic changes to Farol. These include the change to a monetary economy, rather than one where bartering played a major role, and radical changes in employment patterns. These have knock-on effects on the social organization of the village, on family structures, on gender relations, and so on. Traditional gender roles can be either reinforced or threatened by tourism development, partly depending upon the degree of indigenous control over that development (Swain, 1989). For example, the hotel's employment of women may alter their status and role in traditional families. Even though the women are usually engaged in menial tasks and relatively poorly paid, they may become the major (or only) wage-earner in a family (Abbott-Cone, 1995; Pattullo, 1996; Stonich *et al.*, 1995). This kind of break from tradition may well be experienced as liberating by some, but others will regard it as a challenge to the moral 'bedrock' of society. There are many examples where such clashes seem inevitable. Thus tourists who wander round in revealing bathing costumes clash with traditional ideas of modesty, with effects that may be partly responsible for a breakdown of traditional sexual mores: an increase in adultery, and so on. Tourism research shows that this pattern of value conflicts has been repeated in many areas throughout the world (Brown, 1992; Dieke, 1993, 1994).

This leads us to a further qualification of our model. Clearly, tourism as an industry and a social practice is closely associated with modern societies like our own (MacCannell, 1999). Tourism is, after all, as many commentators have remarked, a relatively recent phenomenon. 'To be a tourist is one of the characteristics of the "modern" experience ... before the nineteenth century few people outside the upper classes travelled anywhere to see objects for reasons unconnected with work or business' (Urry, 1997: 4–5). Tourism thus obviously reflects the values of the modern society within which it originates, a society dominated by monetary rather than ethical or aesthetic values. As Marfurt (1997: 174–175) argues, in 'many places an uncritical faith in economic growth ... opened the door wide to mass tourism'. Indeed, tourism has long been defined as a path to economic development precisely because it places a value on objects, landscapes or even people that were previously deemed economically 'worthless' – that is, they may have held a moral, aesthetic or even spiritual but not monetary value. But, as we have seen, economic values are not always predominant in other social formations, nor were they in our own 'pre-modern' past. In medieval Europe, as in many contemporary cultures, there was not such a gulf between the different value spheres, and religious and ethical values played a much more important role in ensuring social solidarity (Durkheim, 1997b). For example, medieval aesthetics was much less autonomous than aesthetics is today; art was much more of a moral endeavour, its iconography subservient to the 'assumptions and orientations of Christian faith' (Bernstein, 1992: 1). For this reason we also have to recognize that the degree of separation (lack of overlap) between the value fields is itself a feature of modernity and that modern societies are characterized by this separation and the growing predominance of monetary evaluations. Thus we can adapt our pictorial model of pre-modern and modern value fields once more, as in Figures 1.3a and 1.3b. These figures indicate that while there is still an interlinking and overlapping relationship between value fields in modern society, economic values have grown in importance while religious or ethical values have become less important than in pre-modern society.

Modernity, then, presents us with a rather paradoxical situation. On the one hand, ethics, aesthetics and economics become increasingly autonomous and self-contained fields (art for art's sake – not for God's sake, as in medieval times), and yet, on the other hand, economics becomes the driving force necessitating changes in aesthetic tastes and ethical values. Ethics may have become more autonomous in the sense that, as we shall see, we have a whole range of distinct discourses and theories about moral values. But ethics has also become increasingly marginal-ized as social life in general has been redefined in terms of the metaphors of the marketplace and hence of economic evaluation.

We shall return to this issue of the commodification of social life and its implications for the ethics of tourism development later, especially in relation to tourism and utilization of wildlife for local development and tourism initiatives in Southern Africa (see Chapter 7). As recent work in tourism studies has begun to emphasize tourism's socio-cultural impacts and the downside to the com-modification of the host–guest relationship (Mathieson and Wall, 1982: 133), so the need to take more than economic values into account in any analysis of tourism development becomes more pressing. Debates about economic valuation and its relationship to ethical, aesthetic and even religious or spiritual values have been absolutely critical to the justification for tourism development in parts of the developing world (Brockington, 2002; Wolmer, 2002). However, the account we have given here of the changing relations between value fields leads directly to some further problems about the nature and importance of ethical values. First, if ethical, aesthetic and economic values have been so intertwined in the past, and still are so in other cultures, and if economic values predominate in modern society and are conveniently quantifiable, then why not ignore ethical values altogether? Why do we not simply subsume ethical and aesthetic issues under economics and deal solely in tangible monetary values? In other words, one could question whether ethical values are really separable from economic values and/or of any real importance. These questions are forms of moral scepticism.

Moral scepticism

So far we have argued that ethics could be envisaged as a relatively autonomous field of values, separate and distinguishable from economic and aesthetic values

(a)

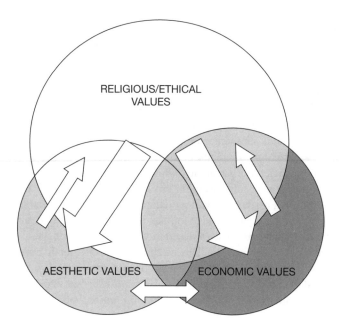

(b)

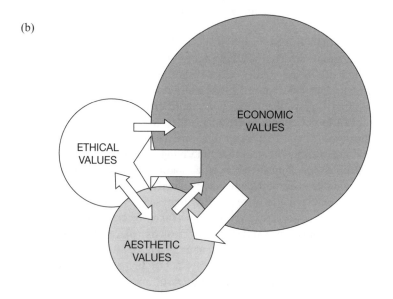

Figure 1.3 Relationship between value fields in (a) pre-modern society and (b) modern society

but interacting with them. The degree of separation between these value spheres, their relative importance, and the permeability of the boundaries between them are indicative of the kind of society concerned. In other words, as Alasdair MacIntyre puts it, ethical values and concepts are 'embodied in and are partially constitutive of forms of social life. [So much so that one] key way in which we may identify one form of social life as distinct from another is by identifying differences in moral concepts' (MacIntyre, 1967: 1). A key way of identifying modernity – that is, the social order in which we all live – is the manner in which economics comes to predominate and to influence many of the areas of life which might previously have been regarded as the realm of ethics or aesthetics. (As future chapters will show, modern societies also employ a range of ethical and aesthetic concepts and practices originally peculiar to them but increasingly exported to other cultures through processes associated with globalization, including tourism. One example of this is how culturally embedded tourist and tourism industry definitions of what constitutes a tropical paradise (a deserted white beach, turquoise water and palm trees) have become globalized. This aesthetic ideal then feeds back into societies in the South, which are then expected to conform through building artificial beaches and planting palm trees (Duffy, 2000a; Edwards, 1996; Lanfant, 1995).)

This predominance of economic evaluation, among other things, leads some people to question the role and status of ethics altogether. If we cannot convince people that ethical values are either real or important, then such issues will inevitably receive short shrift in discussions of the merits and demerits of tourism developments. There are two particularly common forms of moral scepticism that seem to pose a challenge to the taking of ethics seriously. The first, following Williams (1982), we might call amoralism; the second, moral subjectivism (Mackie, 1978).

While amoralists may also be 'immoralists' in André Gide's (1966) sense of the term – that is, people who purposefully set out to break social taboos – they certainly do not have to be. Rather, amoralists claim to be without morals, indifferent to ethics; they do not recognize the authority of any moral codes, usually because they believe that ethics is no more than a subtle form of social coercion employed to get people to do what others want them to. Amoralism argues that ethics is a tissue of illusion that fosters and imposes a façade of altruistic behaviour over and against the underlying reality of people's selfish individualism. People may claim to be concerned for others, but this concern really just masks their own self-interest; the ethically inclined may claim to be promoting the common good but their views really just represent the arbitrary exercise of power by one section of society over another. Take, for example, the case of someone who argues that because mass tourism threatens a place's traditional way of life, it would be wrong for that development to go ahead. From the amoralist's perspective this person uses moral language not out of any genuine concern for that place's inhabitants but in order to benefit themselves in some way, perhaps by keeping the potential 'resort' more exclusive.

The amoralist position was clearly expressed by the political philosopher Thomas Hobbes (1588–1679). For Hobbes, all supposedly altruistic behaviour could be reduced to two very basic 'psychological' motivations: self-interest and the fear of death.[6] Only power enables people to pursue their interests and ensure their own safety. But this search for power inevitably ends in a competitive struggle with others seeking the same end. 'So in the first place, I put for a general inclination of all mankind, a perpetual and restless desire for power after power, that ceaseth only in death' (Hobbes, 1960: 64). Hobbes paints an incredibly bleak picture of human nature and society, with 'every man, against every man' and life dominated by 'continual fear and danger of violent death; and the life of man, solitary, poor, nasty, brutish, and short' (ibid.: 82). His characterization of human motives leaves no room for altruistic behaviour. Indeed, Hobbes argues that all supposedly ethical actions can be reduced to self-interest or fear. 'The affections wherewith men many times bestow their benefits on *strangers*, is not to be called charity, but either *contract* whereby they seek to purchase friendship: or *fear* which maketh them purchase peace' (Hobbes, 1960).[7] In other words, people do 'good' works simply because they think they will profit by them or because they are afraid of doing otherwise. There is no such thing as altruism. The person who returns the wallet dropped on the railway platform does so not because it is 'right' to do so, but because they expect a reward and/or are afraid of being punished for stealing. Box 1.1 provides further illustration of these debates surrounding amoralism.

But is amoralism plausible? Are people always and everywhere selfish, and is all talk of morality nothing more than a smokescreen hiding people's personal interests? If we return to our previous example, then surely, if Muga is right, the new tourist development can only bring the fishermen material benefits, greater wealth, more power? Certainly this argument about tourism as a generator of wealth is often assumed to appeal to self-interest among individuals in host communities. From the Hobbesian point of view, if self-interest and fear of death are the only motives for action, then taking tourists on sunny boat-trips has got to be much less hazardous than fishing in all weathers. Moreover, the extra money the fishermen earn could also pay for better health care and so on. Why, then, do they have moral qualms?

It is always possible for the amoralist to posit ulterior motives and to impugn people's character. Perhaps those who reject Muga's financial overtures are simply holding out for more money or concessions. Perhaps they are afraid of changes that might undermine their prestige within the community and damage their social standing. Perhaps their moral outrage is merely pretence. But is it really plausible always to try to explain away people's moral concerns? Do we have any reason to think that the fishermen's outrage is not genuine, that their qualms would simply disappear if Muga doubled his offer? This seems to be one of many cases where it seems most unlikely that there could be any personal gain from behaving altruistically or sticking to one's moral principles, and yet that is exactly what people do.

Most people do not suddenly start stealing as soon as the coast is clear; most people do not need the threat of legal sanctions to stop them declaring war on their neighbours. Of course, this may just be because they still think it possible that they

Box 1.1   Moral scepticism and the ring of Gyges

In *The Republic* Plato (*c.* 428–347 BC) offers another example of moral scepticism in the form of a story told by Glaucon. Glaucon is an amoralist who claims that we are all led by 'self-interest, the motive all men naturally follow if they are not forcibly restrained by the law' (Plato, 1987: 46). To illustrate this he recounts the myth of Gyges the Lydian, who found a magic ring that would make the wearer invisible at will. Being invisible, Gyges no longer had to fear discovery for any of his 'wrongdoing' and proceeded to infiltrate the palace, 'seduced the queen and with her help attacked and murdered the king and seized the throne' (ibid.: 47). Glaucon argues that everyone would behave similarly in Gyges' position. Indeed,

> if anyone who had the liberty of which we have been speaking neither wronged nor robbed his neighbour, men would think him a most miserable idiot, though of course they would pretend to admire him in public because of their own fear of being robbed.
>
> (ibid.: 47)

Might the annual holiday operate as a spatially and temporally limited version of the ring of Gyges? Some tourists obviously treat holidays as a break from the moral confines of their everyday lives. They indulge in behaviour they might never contemplate at home: having holiday flings, drinking to excess, taking drugs, and so on. This indeed seems part of the attraction of holidays as modern examples of what, in a medieval context, Bakhtin termed the carnivalesque, carnival as something 'liberating from norms of etiquette and decency imposed at other times' (Bakhtin, 1994: 200). There is of course also a much darker side to this in terms of the behaviour of those sex tourists who also utilize the relative anonymity of holiday destinations to engage in practices that are illegal in their home countries. Here, far from providing a respite from, and overturning, dominant forms of social relationships, as Bakhtin argues the carnival did, sex tourism merely exacerbates the commodification of social relations that exemplify modernity (Hall, 1997). People, mainly young women, are treated as no more than commodities to be used and abused. The question remains, though, whether these kinds of behaviour suggest that Glaucon was right, and that given the chance we would all simply seek to satisfy our self-interested desires whatever the cost to others.

will be caught or blamed. But how are we to explain those people who go out of their way to help others, sometimes at considerable costs to themselves and in circumstances where no blame would attach to them for not doing so? Think, for example, of the person who gives up a well-paid job in the private sector to go and work for famine relief or an environmental charity on much lower wages and with more challenging working conditions (see Vaux, 2001).

The sceptic might of course argue that this person would actually get a certain amount of prestige for doing such things, that their motive is to look good in others' eyes. Perhaps they have an angle we can't discern: maybe they hope to get a better job when they return, for example. But why would *we* praise such a person for being moral if we were sure their motives were entirely selfish? And how could any such strategy of hidden self-interest work if there were not already a general approval of self-sacrifice in some circumstances? In fact, if there is no such thing as ethics, why have a moral language at all? Why not simply express everything in terms of self-interest? To counter these arguments, amoralists must engage in more mental gymnastics. They might say we are all being hypocritical when we praise other people's integrity, honesty, and so on. We do not actually appreciate their altruistic or ethical behaviour for its own sake but continue to praise them in the hope that by doing so we can convince them to continue doing things that may one day benefit us. After all, it is better to be safe than sorry, and perhaps one day we might require their 'charity'.

Yet we can think of even more extreme examples where it seems very difficult indeed to see any personal benefit in an altruistic action. Think of Buddhist monks who have doused themselves in petrol and publicly set themselves alight to protest against the Chinese occupation of Tibet. There are many such cases of people laying down their lives for what they regard as a just cause: to save the lives of others or to improve others' living conditions (Vaux, 2001). Such personal sacrifices seem directly to contradict Hobbes's amoralistic thesis, since those people's strong ethical convictions overcome their fear of death, and, as they are dead, there is obviously no way they can *personally* benefit from such an action later.

It seems that the amoralist has to admit, however reluctantly, that self-immolation is not in someone's best interests; that people do engage in activities that entail a degree of genuine self-sacrifice. But, the amoralist argues, we still do not need to admit that ethical values play any part here. The altruistic act is still only an example of that person doing what *they* want to do; it is still a matter of them expressing their own personal preferences rather than their taking an ethical stance. In other words, we do not need to bring ethics into human behaviour but can still explain everything in terms of personal preferences. This is a (much) weaker version of Hobbes's self-interest theory, but even this works only if we assume that a person's wants and preferences are entirely separate from their moral feelings, that morality only ever operates *against* one's personal preferences. But no ethicist has ever suggested that this is the case. The person who lays down their life for a cause may well have *chosen* (preferred) to do so, but this doesn't make their action any less altruistic or ethical. They still do something that goes against their own self-interests in order to express a moral commitment to some 'higher' cause. That commitment is itself an expression of the fact that the person concerned has adopted certain moral standards in such a way that they have become an integral part of their personality, not, as the amoralist suggests, something alien to it. Ethics works precisely because it is embodied in individuals, because, contra Hobbes, it does become part of a 'psychological' profile that is far more complex than he suggests, because our choices are motivated by far more than self-interest.

Amoralists have one last fall-back position in their attempt to undermine the importance of ethical values. They might admit that people do have moral feelings but argue that they do so only because some elements of society have subjected them to a kind of ethical indoctrination. Some people have been successfully persuaded by moral propaganda instilled into them from childhood to behave in ways that are not actually in their own best interests. From this perspective, ethics is thus no more than a form of social manipulation where some are suckered into sacrificing themselves for the benefit of others. Intelligent sceptics: see through this tissue of ethical illusion but usually play along on the off chance that they can manipulate those less aware than themselves to their own advantage. This kind of amoralism is about as cynical as you can get, but before we address this position we should look at just how far we have moved away from the original amoralist position. Despite its aura of scepticism, this kind of amoralism does not conflict at all with MacIntyre's earlier claim that 'moral concepts are embodied in and partly constitutive of forms of social life'. It accepts that moral values are real, that they are not reducible to self-interest, that they have a social origin and that they become part and parcel of people's individual personalities. In other words, it accepts that ethical values are acquired through a process of 'social-ization' whereby people internalize prevailing social norms in the course of their interactions with others and make them their own. This is a position that most sociologists and many ethical philosophers would happily accept, because far from denying the importance of ethics it makes morality a key constituent of a given culture (Joas, 2000). Just as a culture can be identified by its language, its economic system, its forms of knowledge and belief, and so on, all of which are passed on through a process of socialization from generation to generation, so too it is also partly constituted by and dependent upon its own particular system of ethical values.

It makes no sense, then, to see ethics as illusory. The only question becomes one of whether it is possible or profitable to try to escape from the calls of the con-science that to a greater or lesser degree are instilled in us all from childhood onwards. Box 1.2 (p. 24) provides further examination of amoralism. Amoralists would argue that they have ditched morality altogether since it threatened to detract from their autonomy as self-centred individuals who owe nothing to society at large. But no one is ever completely autonomous, and trying to live without ethical values makes about as much sense as trying to live without language. Language too is a cultural product, but we nonetheless use it to think our own unique thoughts and to criticize the thoughts of others. Ironically, without lan-guage we would be much less of a person and much less autonomous; indeed, without language we might not be able to think of ourselves as a person at all. Similarly, we may be socialized into accepting certain moral values, but as we develop they too can be challenged and criticized, and they too add an important extra dimension to our individual character, to our personal integrity and our integrity as a person. This question of the degree to which our moral values might be our creations and property, the degree and circumstances under which we are free to change and exchange our ethical values, leads us directly to the second form of scepticism about ethics, that of 'moral subjectivism'.

Moral subjectivism and economic evaluation

So far, then, we have argued that ethical values are an important constitutive aspect of every society and every (non-sociopathic) individual within that society. But we still are not clear about just what these values are, and this lack of specificity allows another form of moral scepticism to prosper. While the amoralist tries to dissolve the ethical sphere altogether, leaving us only with a realm of values founded on self-interest, the moral subjectivist tries a different strategy. The moral subjectivist admits that ethics do exist and are important, but tries to argue that all such values are really just aspects of one's personal preferences. The subjectivist regards all values, whether economic, aesthetic or ethical, as preferences for different kinds of things: for example, for certain material goods, types of art or landscape, or modes of behaviour respectively. Figure 1.4 provides a way of representing moral subjectivism and the relationship between personal preferences and values.

At first sight this might not seem too contentious a move. It recognizes the same kind of relationship between the value spheres previously set out in Figure 1.3b. It also helps to account for the variation we see in ethical values across and within societies. However, the fact that these value spheres are now all subsumed under the rubric of 'personal preferences' has at least two important implications. First, the values are personal (subjective) in the sense that they are founded in individual *subjects* deemed entirely separate and separable from those around them. Thus the nature of our diagram is changed. Rather than representing the relationship between different values at the level of a society, Figure 1.4 now maps the relationship between values within specific individuals. It is more a map of people's minds than of a whole culture. Each person now becomes the proud owner of a set of economic, aesthetic and ethical values which are theirs and theirs alone, a 'possessive individual' (MacPherson, 1979). (Although clearly the subjectivist does not deny that these values are influenced by others.) Second, reducing these very different kinds of values to something called 'preferences' undermines the relative autonomy of the ethical field. Moral subjectivism reduces ethics, aesthetics and economics to matters of individual taste or desire. Just as I might have a personal preference for chocolate over savoury biscuits, so I might prefer one kind of (moral) behaviour to another (immoral) kind. Thus, to the moral subjectivist the person who says that it is 'wrong' to kill is really just saying that they (personally) do not like murder and would *prefer* people not to go around killing each other. Of course, they are saying it in a moral language, one that is meant to make other people agree with them, but at bottom, for the subjectivist, that is all there is to ethics.

Box 1.2 Amoralism and moral subjectivism in tourism theory

Economic techniques, including cost–benefit analyses, are only one way in which tourism theorists have tried to explain and evaluate the demand for and effects of tourism developments. Unfortunately, many of the most widely employed explanatory models also unwittingly assume or incorporate some degree of amoralism and/or moral subjectivism. That is, they tend to explain host and/or guests' behaviour and values largely in terms of individuals' egoistic and/or subjective responses, tastes and desires rather than in terms of their more complex and culturally specific social and ethical dimensions. For example, Maslow's (1954) famous 'hierarchy of needs' provides a supposedly universal linear rank ordering where those needs relating to 'physiological' functions must be filled before individuals move on to concerns about their 'safety', 'belonging and love', 'esteem', and finally their 'self-actualization'. For those tourism theorists adopting Maslow's work, the key motivation of tourists is 'personal' growth and fulfilment of one kind or another. This is also true of many other accounts, like that of Crandall (1980). From these perspectives tourism becomes something to explain in terms of its potential to satisfy personal desires, to gain individual recognition or status, to keep in shape, think about personal values, and so on. While Maslow does recognize the need for 'ethical' relations and Crandall mentions altruism as a possible motivation for travel, their schemes tend to gloss over the role of societies in producing and maintaining certain sets of ethical norms. These social norms are crucial factors in determining how people actually define their personal motivations and needs. They take the hyper-individualism characteristic of modernity (and dominant within their own models) as a universal norm. The result is a reductive, culturally biased and largely uncritical account of human and tourist behaviour as something almost entirely self-oriented and self-controlled.

While this account might arguably have some validity with reference to egoistic Western tourists, when it is applied generally its limitations become more obvious. Perhaps the key example here is the oft-quoted Doxey's 'irradex', or index of irritation. This venerated typology, described in one recent book as 'one of the most persuasive socially oriented frameworks' (Fennell, 1999: 100), encapsulates in a four-stage model the far from profound insight that locals

> find the demands and impositions of their guests increasingly irritating as time goes by and visitor numbers increase. The 'euphoria' Doxey regards as characterizing the initial stages of tourism develop-ment changes first to 'apathy', then 'annoyance' and finally to outright 'antagonism'. This recapitulation of tourism's social impact as though it

was a failing dinner party is trite almost beyond belief . . . it substitutes pop psychology and a managerial mania for superficial schematization, for serious and critical [social analysis].

(Smith, 2000: 147)

We would argue that the ethical complexities engendered by tourism cannot be accommodated within models which assume that complex social interactions and value differences are explainable in terms of linear extrapolations of individuals' subjective experiences, tastes and desires.

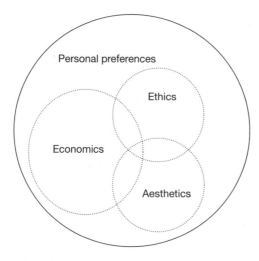

Figure 1.4 Moral subjectivism regards ethical values as one kind of personal preference

Let us return to our tourism example to see what this means. From the perspective of moral subjectivism those people in Farol who make a moral issue out of Muga's plans for tourism development are really just expressing their strong *personal preferences* to be able to continue fishing, maintain their traditions, and so on. When they claim that it is wrong to prostitute themselves and take out boat-loads of tourists, the fishermen are just employing an especially strong language to express their personal disagreement with Muga's plan. But if this is so, then why do they get so upset at the issue being reduced to money? After all, neoclassical economists argue that money too is all about matters of personal preference. The relative values we place on things like a holiday in Ibiza or a trip to the Galápagos Islands depends on how strongly we prefer one over the other. If it costs four times as much to go to the Galápagos, then I need to want to go more than four times as much in order to be willing to pay the difference. Money is thus, say the econo-mists, a measure of the relative strength of my personal preferences for one course of action over another. In addition, it has the advantage that it seems to provide a single linear and numerical scale against which we can compare our incredibly varied tastes and desires – for a new car or stereo, for painting or parachuting, for lazing on the beach or decorating the living room. We must each make choices about what we spend our limited supply of money on, and these monetary choices reflect the relative strength of our personal preferences. So, if economics measures the strength of our personal preferences, and ethics is no more than a matter of personal preference, then economics can surely be used to measure the strength of our ethical feelings. Muga just needs to keep offering the fishermen more and more money until they eventually accept. The point at which they accept indicates the relative depth of their feelings about maintaining their traditions over the goods the money offered can buy them. In this way, the neoclassical economist argues, ethical values are at last made tangible, graspable, manipulable.

Moral subjectivism thus has profound implications for ethics in general and the ethics of tourism development in particular. If ethical values really are just personal preferences, and money can measure the extent of these preferences, then there would seem no reason to get involved in any ethical debate at all. Where conflicts like that at Farol arise, all we need do is employ what economists refer to as a 'cost–benefit analysis', which, as the name suggests, compares the development's potential costs and benefits expressed in monetary values. If the economic benefits outweigh the losses, then the development should proceed; if not, it should not. Of course, as innumerable examples of development attest, there are real difficulties in predicting exactly what the eventual costs and benefits will be. (Think, for example, of the Millennium Dome at Greenwich!) We cannot see into the future and no analysis can hope to be comprehensive since our current knowledge is inevitably incomplete. But at least if the moral subjectivist is right, we may have a way of (quite literally) taking into account people's ethical qualms by placing a monetary value on them.

It is of course relatively rare to encounter a situation like that facing Farol's fishermen, where money is directly offered to effect a change. The economist is happy to speculate about the financial benefits brought by any development, in terms of increased employment, visitor numbers, productivity, etc. But, lacking an actual market, the economist must usually rely on what is termed a 'contingent evaluation' to estimate the values people place on things like clean air, an uninterrupted view, a traditional social practice, and so on (Pearce *et al.*, 1995). This entails employing some kind of survey method to estimate either people's 'willingness to pay' (WTP) to keep things as they are or the level of financial compensation people are 'willing to accept' (WTA) should things change. Once a representative sample of the population has been surveyed, the figures derived from these contingent valuations can then be entered alongside other standard economic data in the cost–benefit analysis.

But things are not so simple. There are a number of problems with the methodological presuppositions of contingent evaluation and cost–benefit analysis that undermine their claim to evaluate either people's personal preferences or, more importantly from our perspective, their ethical concerns. First, this method is profoundly anti-democratic. People's willingness to pay or accept obviously depends on their ability to pay and/or their desperation for money (McAllister, 1980). The poor are typically willing to pay or accept substantially less money in such surveys, and their values therefore count for less than do the values of those financially better off. Many regions and countries where tourism developments are most intrusive are relatively poor (Mowforth and Munt, 1998; Pattullo, 1996). Second, if such surveys were accurate, then one should expect that WTP and WTA methods ought to produce the same figures, but they do not. WTP figures are almost always substantially lower, perhaps because being asked (even hypothetically) to pay out of one's own pocket instils a certain financial 'realism' into this imaginary marketplace, perhaps because there is an obvious limit to what someone can actually pay in relation to what they already have. No doubt partly because of this lower figure, WTP is the preferred method employed when it comes to calculating the deficit side of any cost–benefit analysis. The supporters of a development, who are usually those commissioning the contingent valuation, thereby reduce the 'cost' side of the equation and any potential payments they may be required to make to those who lose out.

There may of course be ways of compensating for such methodological problems – for example, by taking into account the level of people's actual disposable income, and so on. But as we have already seen, there is a much more fundamental problem in trying to calculate monetary equivalents for ethical values, namely, that ethical values are often precisely those which are most resistant to commodification. Indeed, as we saw at the beginning of this chapter, ethical language is frequently employed to denote the limit at which further intrusion of the market into social life becomes unacceptable. The fishermen's traditions, their relationship with the sea, and so on are precisely what they deem *beyond price*. In other words, even if the moral subjectivist is right and ethical values are nothing but personal preferences, then they are very different from economic preferences and not necessarily amenable to the kind of monetarization to which cost–benefit analysis aspires.

This point is illustrated by the fact that 'contingent valuations' are characteristically met with a high proportion of protest bids and/or an outright refusal to recognize the legitimacy of their aims and methods by those surveyed (Sagoff, 1989). Asked to give WTA or WTP figures for preserving a particular landscape, someone's life or a historical tradition, people often give figures of millions of pounds, claim that they are priceless, or deny the relevance of such a question in the first place. Indeed, like Farol's fishermen, people often feel that there is something ethically offensive in even *trying* to put a value on such things. Economists' reaction is usually to treat such objections as irrational (but see Sen, 1977), to exclude the protest bids and ignore those who object to the valuation's presuppositions. But they are wrong to ignore such views, because this adverse reaction to the process of monetarization is indicative of something vital about ethical values. Ethical values are not *quantifiable*, they are not *exchangeable* in the way that giving them a monetary equivalence would imply, and, perhaps most importantly, contra moral subjectivists, they are not just *personal preferences*.

To illustrate this, let us take a hypothetical and somewhat tongue-in-cheek example of a tourist development.[8] Let us imagine that we inhabit a Mediterranean coastal village that has just been purchased by a large corporation with the intention of developing a new 'Zorba the Greek Theme Park'. Utilizing the imagery from the novel and film of the same name, the corporation will flood the village with Anthony Quinn lookalikes, calamari-burger stalls, ouzo cocktail bars and as many cultural stereotypes as possible. They are especially short of authentic-looking ancient women in black mourning dress, and your senile grandmother happens to fit the bill perfectly. For this reason, they offer to take her off your hands and pay you and your relatives a figure to be decided to have her sit in a chair in the marketplace so that tourists can pose to have their photographs taken with her. Some of your relatives believe this is an excellent idea. Your cousin never liked gran anyway, so settles for a nominal €10. Your aunt needs a new refrigerator and so is happy to receive €50 towards it. Uncle Gorgias is not short of cash but always appreciates a business opportunity and negotiates a further €175 for himself. But you think that there is something immoral in this whole process. Greek culture has a tradition of venerating old people, respecting their role in the family and so on, and it is simply wrong to sell one's grandmother, to make a spectacle of her in her old age. You decide that you will not accept any amount of money; your gran's dignity is beyond price, and for the corporation even to offer money in exchange is an affront to everything that is right. Is this stance irrational, as the cost–benefit analyst argues, or does it exemplify the difference between ethical and economic values?

Most people would, we think, agree that a value clash like this cannot be resolved by economic means; that the answer is not, as the economist might suggest, to set up a hypothetical market for aged relatives. To try to do so is to epitomize Oscar Wilde's comment perfectly: one ends up with a price for everything but knowing the value of nothing. It is to try to make one's grandmother an object of exchange, a commodity, rather than treat her as a person in her own right. This is precisely what is immoral, because, as we have seen, one way of understanding the ethical (at least in modernity) is as a 'limit' on the economization of the life-world. Ethics concerns itself with those social relations that cannot and should not be commodified. For example, many arguments against the slave trade were moral arguments: they argued that it was wrong to buy and sell other people, to treat human beings as financial goods. And there are many other areas where we do consider it wrong to make decisions on monetary grounds. Imagine a murder case where the guilt or innocence of a person was to be decided in terms of a cost–benefit analysis; on how much money people would be 'willing to pay' to get an acquittal or a guilty verdict. Any such case would necessarily be a travesty of justice.

If, as the moral subjectivist suggests, ethical values are just personal preferences, then they are obviously preferences of quite a different kind from economic preferences. Either some personal preferences, including ethical values, are so strongly held that they are, by and large, considered untradable, or perhaps the subjectivist is wrong and ethical values are not just personal preferences at all. It is this latter suggestion we want to push a little further.

Most of the philosophical arguments in favour of moral subjectivism are dependent on the implausibility of its alternative, 'moral objectivism'. The moral objectivist holds that there are objective moral standards and/or values that all people can and should recognize, that there are some moral 'facts' of the matter that can be discovered through, for example, rational argument. But the problem is that if we look at actual societies, people clearly disagree about what these standards and values are in a far more radical way than people usually disagree about other kinds of facts. In any case, it is difficult to conceive what kind of moral 'fact' would convince people arguing about the rights and wrongs of, say, sex tourism. Thus people tend to argue that if there are no objective moral values, then they must be subjective matters of personal preference. But from a socio-political perspective this dichotomy between 'objective' and 'subjective' is extraordinarily naïve. As we have seen, the individual subject is always in a large part constituted through their being brought up within particular sets of social relations – for example, as a traditional fisherman. People are socialized from birth into accepting certain values and norms of behaviour and these become part and parcel of their personalities. Only later, if at all, do they challenge these values and norms when, through changing social circumstances, they might find themselves at odds with, or at a sufficient distance to reflect upon, the values they previously accepted unquestioningly. The origins of ethical values are thus 'inter-subjective' and 'relational'; such values are products of, and help reproduce from one generation to the next, certain forms of relations between people and society at large.

Indeed, some classical sociologists, such as Émile Durkheim, argued that ethics is primarily responsible for maintaining 'social solidarity'. While the actual moral values instilled in each society's individual members vary from society to society, their function is always the same: they provide a common normative ground of agreement that ensures that individuals conform to certain predictable kinds of behaviour, thus helping to hold their particular society together. These moral norms are reinforced because those behaviours that conform to social expectations are deemed praiseworthy and those that do not are condemned and punished. This sociological perspective which (like Alasdair MacIntyre) sees ethical values as partly constitutive of social life also helps to explain both why ethical values are so deeply held and why they are so easily affected by social change of the kind affecting Farol.

Durkheim also suggests why ethical values can and should be distinguished from economic values. He argues that every society is divided into what he terms the 'sacred' and the 'profane'. The profane is associated with everyday objects that can be appropriated and transformed by individuals without breaking any social taboos. By contrast, the sacred is associated with those things, places or behaviours that transcend (go beyond) the remit of any individual, that are deemed beyond any individual appropriation. This is because Durkheim regards the sacred as a symbolic expression of the social order as a whole, as an inter-subjective product and reflection of those shared values that hold any society together. For Durkheim, societies surreptitiously make their gods and their ethics in their own image. The religious and the ethical were originally one, and both partake of the sacred. Both get their power and their mystique from society itself, because they are constitutive of and help constitute the underlying social order. Sacrilegious or immoral behaviour is thus a transgression of and a challenge to the social order itself (which again explains why there are social sanctions against breaking moral codes). In the words of Steven Lukes (1988: 26):

> Thus on the one hand, there is the sacred – 'elaborated by a collectivity', hypostasizing collective forces, fusing individual *consciences* 'into communion', imposing respect and love, transferring 'society into us' and connecting 'us with something surpassing us'. On the other hand is the profane – expressing 'our organisms and the objects to which they are most directly related', and relating to men's [*sic*] ordinary life, which is seen as involving 'daily personal preoccupations', 'private existence' and 'egoistic passions'.

Although the extent and universality of Durkheim's overly dualistic conception is open to challenge (Lukes, 1988; Smith, 2001), it does provide a basic rationale for understanding cultural variations in moral codes and values. It also explains why ethical values are not reducible to matters of subjective personal preference – because they are social, not individual, products – and why people are clearly discomfited when economics (the profane) breaks social taboos and intrudes into issues regarded as the preserve of ethics (the sacred).

We shall return to some of these issues later, but for now we hope that this chapter has established a few key points:

1 That there are numerous kinds of values including ethical, aesthetic and economic values.
2 That ethical values are important and not, as the amoralist claims, superfluous and/or reducible to matters of self-interest. Nor, contra the moral subjectivist, are ethical values reducible to personal preferences. If the arguments presented against these two forms of moral scepticism are accepted, then ethical debates are in no way reducible to debates about economic interests.
3 That ethics are partly constitutive of, and inextricably involved with, particular forms of social life and cannot be understood without some recourse to those forms of life. This also means that values vary within and between different cultures.
4 That tourism is clearly associated with a particularly modern form of social life and is therefore likely to reflect modernity's own values. In modernity, ethics and aesthetics are expelled from core social activities and marginalized in key decision-making processes (Figure 1.3b). This is part of a process of economization or commodification in which economic exchange values gradually come to predominate over all other aspects of modern life. Tourism developments, like modernity, are largely led by economic considerations.

It is our contention, then, following on from these points, that the recent turn to ethics in tourism literature should be seen in context as a response to, and sometimes as a part of, a growing resistance to the processes of commodification entailed by tourism developments. Ethical issues arise where this process is regarded as having gone too far or too quickly, where its social and environmental repercussions have been sidelined in the name of profits. The question remains to what extent ethics can and should operate as a limiting factor in these developments and whether tourism can become ethical. We will raise these issues first in the context of the cultural (and moral) variations within and between cultures and the various ethical frameworks that have been devised in order to attempt to mediate and adjudicate between differing ethical values.

2 The virtues of travel and the virtuous traveller

'Self-discovery through a complex and sometimes arduous search for an Absolute Other is a basic theme of our civilization,' says MacCannell (1999: 5) in his now classic sociological text *The Tourist*. Tourism might, then, at least in part, be regarded as an expression of this socially conditioned desire to encounter that which seems 'exotic', 'different' or 'new' to us (see Chapter 6). Western culture has often regarded this search as something virtuous in itself – that is, as something that is quite literally 'character forming' in a positive sense. We become 'better' people for our experiences of different lands and cultures. Gazing on the Other in tourism can be part of a process of self-discovery and performance of the self, whereby tourists can define their self-identity in contradistinction to the Others they have come to gaze upon (Galani-Moutafi, 2000). But the search for Otherness also brings individuals and the tourist industry into contact with the values, desires and needs of those inhabiting other cultures. Tourism thus presents a number of ethical questions about encountering cultural differences and about becoming a more 'cultured' individual, about the journey to other lands and journeys of self-discovery. These form the topics of this chapter.

Ethics and cultural difference

One of the main conclusions of the previous chapter was that ethical values are key components of particular societies because they help structure the patterns of behaviour that (at least in part) constitute that society. Durkheim (1968) suggested that ethical values are vital precisely because they function to hold society together; that our respect for common (sacred) values is the bedrock on which a society is founded. He thought that this contribution of morality to 'social solidarity' would be most obvious if we examined simple forms of society with little or no social differentiation. Here the lack of any extensive 'division of labour' and the fact that each person is capable of fulfilling much the same tasks as any other means that everyone is likely to see and value the world in the same way. In short, the members of such a supposedly 'primitive' society stick together because their social roles and values are almost identical.[1] For this reason, Durkheim thought that the 'elementary' religious/ethical aspects of a society took the form of what he termed a *conscience collective*. This term was meant to imply both a collective *consciousness* in the sense of a shared understanding of the social world and a collective *conscience* in terms of shared moral values.

In practical terms this means that we would expect to find little *internal* disagreement within such societies over moral issues. Modern societies are, however, obviously quite different in the sense that they involve an extensive division of labour. People occupy quite specialized positions, as doctors, politicians, hoteliers, electricians, tour operators, and so on, all of which can give rise to quite different perspectives about what is right and wrong. The fact that our individual values can differ so widely means that the ethical conformity associated with an extensive *conscience collective* can no longer exist. Ethical disagreements occur frequently and can often seem quite divisive: think of the fundamental disagreements that exist over issues like capital punishment, abortion, euthanasia, and so on.

But even here, in modern societies, Durkheim argues, ethics still acts to foster a form of 'social solidarity'. Ironically, the fact that we all have much more individualistic perspectives also means that we all come to respect an ethic of individualism, to share a view about the value of individual autonomy itself. This 'cult' of individualism is the grounds of a new (though noticeably weaker) form of collective consciousness and conscience that is enshrined in our (democratic) political procedures and our laws. Durkheim also claims that the very fact that everyone has a specialized task also means that we are all increasingly reliant on each other to provide the requirements of everyday life, that we can no longer do everything for ourselves. For example, the tour operator cannot succeed without hoteliers, caterers, and so on all fulfilling their own very different functions. This need to rely upon and respect the different roles and values that other people have gives rise to what Durkheim refers to as 'organic solidarity'. Like the human body, modern society might be regarded as being composed of many specialized 'organs', the social equivalents of the heart, liver, lungs, etc., each with their own specific and irreplaceable function but each absolutely interdependent upon all the other organs for their survival. The functional differentiation of the body is analogous to the functional differentiation caused by the division of labour in modern societies. Thus, in contrast to the 'mechanical solidarity' of 'primitive' cultures where everyone's roles and values are so nearly identical that they might almost have been machine produced, modern societies give rise to an ethics of mutual interdependence, an ethics based on the difference rather than the similarity between us.

While many might argue that Durkheim exaggerates the gap between modern and pre-modern societies, his work remains useful because it develops a quite sophisticated account of the changing relationships between ethics and society. It also serves to highlight a number of vitally important issues underpinning all attempts to get to grips with the existence of disparate ethical evaluations in different social circumstances. These might be summarized under three headings:

1   the problem of moral relativism;
2   the nature of the relationship between social roles and ethical values;
3   the problem of how to transcend or negotiate ethical differences.

This chapter will address each of these issues in turn, but they will resurface again in later chapters as we further examine the ethical dimensions of tourism development in specific contexts.

Moral relativism

As we have seen, Durkheim develops what might be described as a 'normative' account of ethics. That is, he thinks ethics engenders social solidarity by setting out the social norms to which members of any given culture are expected to conform. These norms, whether explicitly stated or implicitly accepted, serve to prescribe the boundaries of acceptable social behaviour, and failure to comply with them risks punishment. For example, male homosexuality is illegal in Sri Lanka (Niven *et al.*, 1999: 83) and can incur a jail sentence in Malawi. These codes of conduct vary from society to society, and we can only understand how each of these diverse forms of ethical evaluation functions by examining its actual social context. But emphasizing the close relationship between the forms taken by a society and its ethical values also seems to suggest that all ethical values are socially relative. Many philosophers worry that this kind of sociological account of ethics almost inevitably leads to a full-blown 'moral relativism' that regards all values as equally valid and undermines any attempt at ethical criticism.

We shall argue that this is not necessarily the case, but it is still important to understand why some people are concerned about moral relativism. Indeed, the claims made for and against moral relativism are especially relevant to an area like tourism development, where ethical disagreements often arise from an apparent incompatibility between the competing cultural values of 'hosts' and 'guests'. For example, tourism can be perceived as positive and negative in communities where sharing and preserving local culture could be seen as conflicting goals (Besculides *et al.*, 2002; see also Lindberg *et al.*, 2001). Tourism can be regarded as a real threat to traditional, religious or cultural values even when it is clear that the local community engages with the tourism industry and derives significant economic benefits from it (Joseph and Kavoori, 2001; Kousis, 2000). It is important to understand what moral relativism is and why people might hold this position. Moral relativism can take several forms, but the arguments in its favour usually incorporate the following basic claims into an argumentative structure:

1   The fact of moral and cultural diversity: there are many different cultures in the world, each with its own set of ethical values.
2   The functional inter-relationship between ethics and society: ethical values are an integral part of each society and cannot be understood outside that context.
3   The social relativity of evaluation: what counts as right or wrong depends upon which culture you belong to.
4   The critique of objectivity: if all ethical values are socially relative, there cannot be any *objective* ethical values.
5   The importance of self-reflexivity: our own ethical values too must be socially

relative; they cannot be objective, nor can they provide us with a privileged perspective on what is right and wrong.

6   The equality of ethical evaluations: since there are no objective criteria by which to judge moral values, they must all be treated as equally valid. We should be respectful and tolerant of other people's moral codes, however different they may be from our own.

From the perspective of tourism the moral relativist's conclusions seem almost reassuringly clear. The beliefs and values of the culture we are travelling to might be very different to our own but they are equally valid. Since all values are socially relative, there is no way of deciding which of us is objectively right, and we must respect the host's ethical evaluations and, so far as possible, conform to their customs – 'when in Rome, do as the Romans do'.

But is it this simple? While many of the claims made by moral relativists certainly appear plausible, this line of argument may not be as straightforward as it initially seems. A sociological account of ethics such as Durkheim's obviously accepts stages 1–3 almost as read: cultures are morally diverse, ethical values do play a role in creating and maintaining cultural differences, and opinions as to right and wrong do vary between cultures. However, the situation is actually much more complex than the relativist suggests, partly because this argument treats societies as easily distinguished and securely bounded entities. In reality, societies rarely if ever exist in complete isolation from other societies; they share some aspects of other societies' forms of life, they may have a common history, live in similar geographic locations, trade with each other, hold similar religious views, and so on. This is especially true of the modern world, where, MacCannell (1999: 77) says, 'otherness is mainly fictional as modernity expands and draws every group, class, nation and nature itself into a single set of relations'. If this is so, then we have to look much wider than the internal constitution of a society to understand its ethical values, and we might also expect to find some shared values across societal boundaries. Similarly, there may be extensive divisions in moral attitudes within a society, meaning that some people's values have more in common with those of some others outside rather than within their own culture. This does not necessarily undermine Durkheim's account; it simply means that, contra an extreme moral relativism, we can usually expect to find some common ethical ground between 'hosts' and 'guests' about matters of right and wrong.

The question of moral relativism arises frequently in discussions of ethical behaviour among tourists. For example, consider the problems that might be encountered by a feminist travelling in a society where women appear to be treated unfairly. It is relatively simple to accept that social and religious norms mean that a woman must cover up when entering a mosque as part of a tour. However, what about when a woman is travelling in a conservative Muslim country – indeed, 'In most Third World countries – whether Christian, Buddhist, Hindu or Muslim – fashion is fairly conservative' (Wheat, 2001). Does the Western woman accept that she must wear long skirts and cover her head? Or if she agrees with some Islamic feminists that wearing such clothing represses women, does she then decide to wear Western clothing and not to cover her head, risking offending both women and men in the society she is visiting? (See Afshar, 1998; Craik, 1995; El Sadaawi, 1997; Enloe, 1990.)

Some philosophers have tried to counter moral relativism by arguing that the common ground within and between cultures is actually very extensive; indeed, that there is actually universal moral agreement over certain issues, or at least would be if everyone thought through the issues in a sufficiently rational manner. In other words, they would disagree with stage 4 of the above argument and claim that despite the apparent ethical diversity between cultures there are *objective* moral values, values that are not relative to any particular society but either are, or could be, shared by all. But, as Chapter 1 argued, this argument for moral objectivism seems difficult to sustain given the real divergence of opinions and the intense disagreement about exactly what might count as an objective value. For example, although many societies might superficially seem to agree that murder is wrong, on more detailed investigation they almost always allow killing to take place in particular circumstances: war, old age, heresy, illness, and so on – circumstances which themselves vary considerably between and within cultures.

Leaving aside the arguments about moral objectivism, there are other reasons for rejecting moral relativism's rather simplistic conclusions, especially where tourism is concerned. Even if we are able, as step 5 suggests, to be self-reflexive and thereby come to recognize that our values are also socially relative and cannot be established as valid in any objective fashion, this does not mean that we can simply decide to abandon them. The very fact that they have been such an important and integral part of our social environment and our own self-identity means that they cannot be compromised without affecting our relations to our own society and our personal integrity. Being able to go along with and respect the values of another social group simply may not be an option where it conflicts with our core beliefs. Should we respect local views when we might consider them homophobic, racist or otherwise discriminatory? In any event, there is no reason why moral relativism should actually imply such tolerance, because, as Bernard Williams points out, the moral disagreements between parties might also include disagreements 'about their attitude to other moral outlooks' (1982: 37). In other words, a society might think the right thing to do when confronted with a different set of cultural values is to ignore or subdue them, not tolerate them. If we are going to be fully self-reflexive, then we had better realize that tolerance too is a socially relative virtue.

Vegetarians often face complex ethical issues when they engage in tourism to the South. An individual may be deeply committed to the idea that it is inhumane and morally wrong to eat meat. However, they may then find themselves in a society where food is scarce and there is no local understanding or acceptance of vegetarianism. Should the vegetarian visitor compromise their deeply held moral beliefs and eat meat? In this scenario they might have to weigh up the need to be culturally sensitive, recognizing that in a situation of food scarcity to refuse any food would be regarded as an insult, against risking alienating their hosts through explaining their vegetarian beliefs. There is no easy answer to this from the point of view of the tourist or the local community. The cultures that tourists visit are not unchanging and/or necessarily resistant to new ideas, but there is a complex power relationship between the Southern host and the Northern guest which further complicates the issue (Joseph and Kavoori, 2001).

Rejecting moral relativism's conclusions does not undermine the importance of trying to understand the many and complex ways in which ethical values are related to particular social contexts. It should, however, make us wary of drawing simplistic conclusions about how to accommodate ethical differences. We cannot simply ignore what others think about right and wrong, because to do so will cause offence. The tourists entering the church of Santo Tomás in Chichecastenango, Guatemala (see Plate 2.1), commit a form of sacrilege in the eyes of their Quiché hosts when they ignore requests not to take photographs inside this sacred space. To ignore such requests is obviously not so blatant a disregard for others' values as to forcibly impose one's ideas on them – as the original conquistadores who constructed the church over the site of a Quiché sacred rock did – but both clearly involve a lack of cross-cultural respect. On the other hand, simply tolerating what goes on in other places can be just as problematic. At the entrance to Chichecastenango is a sign that reads *El Ejército es su Amigo* – 'The Army is your Friend' (Lingis, 1995: 251). This same army was recently responsible for the torture and extermination of tens of thousands of the local population. Do not tourists have some sort of responsibility to understand, and have opinions about, the often morally problematic background of the culture they visit? In fact, such examples clearly show us the complex ways in which even very different and distant cultures are not isolated entities but are caught up in shared histories,

Plate 2.1 The church of Santo Tomás in Chichecastenango, Guatemala, 1994

economies and politics. In this case, the question of making the most of one's photo-opportunities cannot be seen in isolation from the original greed-driven incursions of the conquistadores or the Cold War politics that funded a repressive military in the name of anti-communism (Shoman, 1995; Weinberg, 1991).

Social roles and virtues

The question of moral relativism began from the simple idea that different kinds of society have different values, leading us to focus on cross-cultural differences. However, as we have seen, Durkheim also distinguished between so-called primitive societies governed by a collective consciousness/conscience and modern societies characterized by quite extensive *internal* value differences. These internal differences were, Durkheim claimed, a product of people adopting different moral perspectives as a consequence of the more extensive division of labour. The more specialized the tasks we are engaged in become, the more disparate our social roles and the greater the potential for ethical differences between us. However, our own practical experiences tell us that the degree of organic interdependence of people in modern societies is not necessarily enough to ensure that ethical differences are always tolerated or that everything runs smoothly.

It seems undeniable that ethical tensions are widespread in modern society. Alasdair MacIntyre claims that 'the most striking feature of contemporary moral utterance is that so much of it is used to express disagreements'; there seems to be no way of 'securing moral agreement in our culture' (1993: 7). Indeed, Chapter 1 argued that it is precisely the presence of such seemingly interminable ethical disagreements that gives credence to various forms of moral scepticism. If there are no externally accepted principles, criteria or values against which to measure an individual's behaviour, then it is easy for the moral subjectivist to argue that ethical values must be no more than expressions of personal taste. Yet some division of labour and degree of value difference seems inevitable. There are few, if any, societies that perfectly exemplify Durkheim's mechanical solidarity. If we look back at the recorded history of cultures that preceded our own, we find that there have always been people occupying specialized roles as priests, rulers, soldiers, and so on. What, then, are the differences between these societies and our own? What other factors affect the relationship between individuals, their social role and ethical values in ways that might account for the particular prevalence of moral disagreements in modernity?

One possible factor is the relative stability of a culture's social relations over time. MacIntyre argues that where cultures (and the division of labour) remain the same over generations, the different roles in society – priest, nobleman, farmer, and so on – come to be closely associated with their own set of role-based moral virtues. Thus in ancient Greece, around the time Homer's *Iliad* and *Odyssey* were composed (*c.* 700 BC), a good nobleman would be expected to be 'brave', 'skilful and successful in war' and 'wealthy' (MacIntyre, 1967: 6). Homer uses the term 'good' (*agathos*) in a very specific way. It means nothing more nor less than to have the character traits, the virtues, necessary to fulfil one's role successfully.

To be 'good' was synonymous with being good at one's job (social role). A 'good' nobleman simply could not be unsuccessful in war or poor, since he would have lacked important virtues associated with his role. These virtues thus provide socially accepted criteria against which the moral worth of the person concerned can be measured. Of course, the particular moral character traits required varied from role to role. 'Fidelity', for example, was a virtue expected of the nobleman's wife but not of the nobleman himself. Thus Homer praises Penelope for putting off her suitors during the many years of Odysseus's journey home from Troy but attaches no blame to Odysseus himself for his indiscretions on the way!

This is not (just) a case of double standards, but an active espousal of multiple standards. Every social position comes with its own set of expectations. Perhaps one way of picturing this is to think of those children's toys that involve hammering different-shaped blocks of wood or plastic into appropriately shaped holes cut into a solid frame. We might think of the framework as representing the social structure of the society with a number of different social roles (holes) needing to be filled. What morality requires is that there should be a good fit between the individual blocks of wood – that is, the individual person, in our analogy – and the social roles available to them. Each role (hole) requires the individual (block) that will fill it to have different characters: to be triangular, hexagonal, circular, etc. These characteristics are the equivalent of the character traits, the virtues, that the individual should exhibit in order to fulfil their role. The more perfectly you fit the role assigned to you, the more virtuous you are. Being a good nobleman would require you to have completely different character traits (a different shape) as compared with being a good prophet, farmer, and so on. It is no good trying to fit a square block in a round hole. Here, then, we have a situation with a quite extensive division of labour associated with different moral evaluations but where social stability (a solid frame) ensures that there are external criteria attached to specific positions within a society. Figure 2.1 illustrates this.

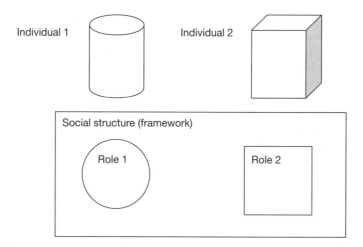

Figure 2.1 'Playing' with social roles and individual values

This kind of association between roles and virtues is not of course confined to ancient societies, but it does require a fairly settled social order and/or clearly distinguished social roles. This is why talk of virtues still seems to make some sense in the case of members of particular professions – for example, nursing, where one's ability to act professionally might be said to depend upon fostering specific character traits. Florence Nightingale, that most famous of nurses, argued that a *good* nurse needed virtues such as 'truthfulness, obedience, punctuality, observation, sobriety, honesty, quietness, devotion, tact, loyalty, sympathy, humility' – a list that one author, somewhat contentiously, claims 'needs little or no revision' today, more than a hundred years later (Sellman, 1997). It is not difficult to see that an appropriate list of virtues might also inform the codes of conduct for various professions in the tourist industry, like tour operators or holiday representatives, though the list might differ somewhat from that deemed suitable for nursing. Indeed, it might even differ across different tour operators, since the ideal holiday representative for a company specializing in holidays for the over-sixties would have to be conspicuously different from one looking after the clients of Club 18–30! The roles played by tourism professionals may also be shaped or even defined by institutions or sets of values outside the tourism industry itself. For example, tour guides in Suharto's New Order regime in Indonesia were required to present the government in a favourable light to foreign visitors (Dahles, 2002). Thus government intervention can mean that real restrictions are placed on guides and can be an important determinant of the roles performed by guides and other tourism professionals.

Perhaps, however, there may be some common features across a variety of roles taken by tourism professionals. Many authors have pointed out that those at the sharp end of the tourism/hospitality industries, such as holiday representatives, waiters, hotel receptionists, and so on, are expected to adopt a particular kind of role in relation to their customers (Urry, 1997). As Marshall (1986: 41) points out, they must 'cater for' the tourists, smile, exchange pleasantries, and generally learn to give a performance as someone who is friendly, approachable and willing to help (see also Warren, 1999). Hochschild (1983), who studied the way in which the cabin staff of airlines were trained, refers to this as a kind of 'emotional labour' in which the whole point is to foster a particular kind of emotional response in customers in order to put them at their ease. The more strongly engaged in this social performance employees become, the more effective they are at selling an idealized personality *and* selling the company's services. The way that a hotel concierge has to appear as smiling and welcoming and simultaneously display their profession-alism indicates the ways that tourism workers are encouraged to play specific roles to satisfy the needs of visitors and the tourism industry. The employee engages in a form of 'method acting', adapting their demeanour to fit with the employee's and tourist's expectations of 'virtuous behaviour' even though these displays of emotion may conflict with their own personal feelings. Failure to behave in the expected manner can lead to dissonance and conflict, as in the case of street vendors and hawkers, who are often characterized by tourists (rightly or wrongly) as harassing (de Albuquerque and McElroy, 2001). Emotional labour takes many forms, some of which are merely innocuous but irritating, like the forced smiles which accompany stock phrases like 'have a nice day'. Others are more extreme, as in the case of the beachboys of the Gambia, who must perform a very different role through their provision of sexual services for women tourists (Brown, 1992; Hall, 1997). This management of emotions thus raises further ethical issues about the 'authenticity' of tourist experiences (see Chapter 6) and about the imposition of patterns of behaviour on those who must sell their emotional labour as a commodity.

In this rather limited sense, qualities like 'friendliness', 'conviviality', 'courtesy', 'helpfulness' and so on might be regarded as appropriate virtues for those engaged in certain roles related to tourism. But thinking about virtues in this way illustrates the differences between the place of the virtues in Homeric Greece and in modern society. The virtues associated with being a Greek nobleman formed an integral and unifying part of the narrative of that person's life, and the nobleman's ability to exhibit these character traits determined his moral standing in the eyes of society as a whole. In ancient Greece you had a 'job' (role) for life and your life was that job. But this is hardly true of the travel representative or flight attendant. In modern societies our social roles change over time, there is considerably more social mobility, and we may have several quite unrelated jobs over the course of our lives. People also simultaneously occupy roles that have little or nothing to do with their actual employment – for example, as a mother *and* a manager, a consumer *and* a consultant. These roles may require us to exhibit very different sets of 'virtues', from the cut-throat competitiveness of the commercial operator to the sympathetic care of the parent.

Within the tourism arena it is clear that tour guides play multiple roles to satisfy the desire of tourists which are quite separate from their roles at home or in social settings (Duffy, 2002). For example, tour guides in Belize feel the need to restrict their conversations with tourists to certain topics that will enhance their holidays and ensure that they leave a tour with the impression that they have enjoyed themselves. Consequently, tour guides very often talk to their clients about activities they choose to engage in while on vacation, such as scuba diving, sport fishing, dancing and drinking (see Chapter 6). However, once the tourists have gone, young male tour guides often turn their conversation to more Westernized and globalized topics, such as what the latest Nike trainers look like, or whether their friends have seen the most recent Hollywood release. Guides are very aware that trainers and films do not conform to the tourist's idea of what local people (specifically guides) 'should be like'. Instead, they tend to accept and perform the role of the guide who is only interested in fishing and scuba diving. The concept of tour guides as mediators of local cultures fails to capture these complexities. The tour guides choose, or feel compelled, to present only certain aspects of their complex culture, which draws on local tradition but is clearly located within broader processes of globalization. Tourist expectations of a homogeneous, pre-modern society flatten out all reference to globalization and cultural hybridity. For example, Moreno and Littrel (2001), in their study of Guatemalan textile retailers, found that the retailers conceptualized tradition in ways that assisted them in selling their products to tourists.

In short, our lives lack the one-to-one correspondence between the single social role and its associated virtues found in Homeric Greece. The almost infinite variety of combinations between differing social roles, together with the rate at which these roles appear, disappear and change, means that there is no time for stable systems of moral virtues to emerge and gain widespread social acceptance. We live, in Zygmunt Bauman's terms, a 'life in fragments', pulled in different directions by the 'overall tendency to dismantle, deregulate, dissipate the once solid and lasting frames in which life-concerns and efforts of most individuals were inscribed' (1995: 265). In modernity one's social role is often 'merely temporary and may disappear virtually without notice' (ibid.), making everything, including virtues, seem much more contingent.

This fragmentation also means that a gap develops between our self-identity and our social role(s). Our own private values are not necessarily those associated with our mode of employment. Put bluntly, the flight attendant's smile is usually a performance paid for by company cash rather than an expression of an underlying personality (for further discussion, see Warren, 1999). Although it is usually in the interests of employers, employees and customers alike to make the performance seem as genuine as possible, the emotional wage-labour of those employed in service industries might be regarded in Marx's terms as a form of 'alienated' labour. By that we mean that the very fact that this emotional labour is owned and managed by others in order to make money means that it becomes estranged and disassociated from the interests, values and life of the actual employee. And although there are examples of so-called 'company men' [*sic*] whose whole life is entirely devoted to fulfilling their corporate role, in Western culture at least, these people are often judged to be morally deficient by those outside (or even inside) the firm rather than virtuous or praiseworthy.[2] They are often referred to as 'workaholics', blamed for disrupting family life, or exhorted to 'get a life' (outside of work). The fact that we recognize this gap between modern individuals' values and their social role(s) also means that it is usually deemed inappropriate to pass moral judgement on the employee in the same way that Homer passed judgement on his historic characters. We would be unlikely to decide whether or not someone was a 'good' person on the basis of his or her ability to produce a party atmosphere in the hotel or serve drinks with a smile. A poor 'performance' by staff is just that, a failure to *act* their part, and is more likely to reflect badly on the company, which is regarded as not having properly trained or managed its employees to live up to expectations.

Where does this leave the idea of virtues in modern societies? Initially, at least, it seems that their influence is much reduced. The separation of the individual's moral standing from the job they perform and the qualities associated with that job means that the virtues often become little more than items on a wish list in company job specifications. 'Wanted: lively, outgoing person with an ability to communicate with others and manage their time effectively.' Such qualities are as much about having acquired 'transferable skills' as they are about the indelible character traits more traditionally associated with virtues. It seems that to some degree MacIntyre (1993) is right to argue that we live in a society 'after virtue', where the erasure of communally agreed standards of behaviour throws the responsibility for making moral decisions back on the individual, an individual now left with precious little external guidance. (A situation which, as previously suggested, explains the prevalence and power of arguments for moral subjectivism.) Unsurprisingly, many people, including MacIntyre himself, find this modern condition daunting and seek salvation in a return to shared social practices and shared moral communities. But as Bauman (1995: 277) points out, tightly knit communities may liberate the individual from the responsibility of making their own moral choices only at the cost of subjecting them to an oppressive regime where 'independence is frowned upon, dissent hounded down, disloyalty persecuted . . . [where] the craved-for cosiness of belonging is offered at the price of unfreedom'.

We will return to the relationship between the individual and community later when we discuss issues of social justice in Chapter 5. First, however, we will examine other possible ways in which we might be able to retain a conception of virtue without necessarily linking it to specific predefined social roles. This will lead us directly into a discussion of the virtues of travelling as a mode of character formation.

Virtues: on being human

Modern societies are not the first to have had to face the dissolution of traditional links between social roles and ethical values. Just a few hundred years after Homer, those Greeks inhabiting city-states like Athens found themselves in a similar position. Where much of the Mediterranean seems to have been a place of mystery and rumour to Odysseus and his storm-tossed crew, the Athenians were in regular, close contact and trade with peoples from all around the Mediterranean and beyond. Something of the variety of peoples and influences can be seen in a passage from Athenaeus, who cites some lines from the comic dramatist Antiphanes:

> Cooks from Elis, cauldrons from Argos,
> Wine from Phlius, coverlets from Corinth,
> Fish from Sicyon, flute-girls from Aegion,
> Cheese from Sicily . . .
> Perfume from Athens, eels from Boeotia, . . .
> From Cyrene stalks of silphium and hides of oxen,
> From the Hellespont mackerel and salted fish of all kinds,
> From Thessaly puddings and ribs of beef,
> From Sitalces, an itch to bring the Spartans up to scratch,
> From Perdiccas, fleets of ships with cargoes of lies.
> (Athenaeus in Ferguson and Chisholm, 1978: 87)

And so the list goes on for many more lines. Each trading relation brought more than new goods: it encouraged the growth of an ever more complex set of social relations that included new and rapidly evolving social roles. It also brought the Athenians into contact with cultures employing a wide variety of very different moral values, which, together with Athens's fledgling democracy and increasingly varied, educated and critical culture, posed a serious challenge to traditional roles and virtues. This is a scenario that is especially familiar in areas that have experienced tourism development. Interaction with other cultures is an almost inescapable element of tourism, and is particularly relevant in tourism to the South by travellers from the North (Galani-Moutafi, 2000; Joseph and Kavoori, 2001).

The Greeks came up with some ingenious answers in order to retain the idea of the virtues and a virtuous life. Some, like the Sophists, a school of itinerant teachers who made their living from training the wealthy of various cities in political and legal rhetoric, simply adopted a full-scale moral relativism. The techniques of persuasive argumentation they taught could be easily adapted to fit with the particular virtues recognized by the cultures in which they found themselves. But this willingness to concentrate on the form and techniques of argument at the expense of any substantive moral content ensured that many of their contemporaries regarded their teachings with suspicion, a suspicion still reflected in our own use of the term 'sophistry'.

Plato, often regarded as the founder of modern Western philosophy, had quite a different solution. He argued that with proper philosophical training it was possible to see beyond the inconsistencies that will inevitably characterize common opinions in an unstable world ruled by change and decay. Through the exercise of reason and argument philosophers are able to gain real knowledge of the essence of what constitutes 'the good', knowledge they can then use as a template by which to measure the actions of their contemporaries. This solution, however, requires one to believe that there is an independently existing idea of 'the good' out there waiting to be discovered and that philosophers will all necessarily agree on exactly what form this takes. In other words Plato's theory suffers from many of the faults associated with moral objectivism that were set out in Chapter 1.

Perhaps the most important and lasting contribution to virtue theory was that of Plato's pupil (and Alexander the Great's tutor) Aristotle (384–322 BC). Aristotle sought to resolve the problem of moral diversity by turning to a model of human nature. For Aristotle a human being was a rational creature that could be defined as a *zoon politikon* – that is, a political or social animal. Each human strives to attain a fulfilled, flourishing and happy life (the term Aristotle used was *eudaimonia*). The virtues are therefore those character traits that allow us to flourish as humans – that is, to achieve and enjoy our full potential as rational and social animals. They are those characteristics that we should aim towards in order to further our well-being and happiness in a community of fellow humans, characteristics that, Aristotle claims, are the same for all people in all cultures. In this way he seeks to overcome the problem of moral relativism without resorting, as Plato did, to rather bizarre metaphysical ideas. For Aristotle the proper human life is an ongoing project that seeks to fulfil our full human potential through our striving to act according to the virtues and to make virtuous choices.

How, then, do we recognize the virtues? What kind of character traits would help social animals flourish? Aristotle argues that our own well-being and that of those around us require us to choose a reasonable and reasoned path between any extremes of behaviour. In this way we fulfil our duties to others while at the same time ensuring that we utilize our full human potential as rational and social animals. This is often referred to as the doctrine of the 'golden mean' and is best illustrated by examples. A virtue such as 'courage' is, says Aristotle, a kind of sensible (golden) mean between two extremes, those of cowardice and rashness. Where the coward is 'a despondent sort of person' (Aristotle, 1986: 129) deficient in their ability to face danger, the rash person is excessively fearless because their response to danger is not properly tempered by reason. Anyone who is 'afraid of nothing – not even the earthquake . . . would be a maniac or insensate' (ibid.). The courageous person certainly feels fear in appropriate circumstances but they face that fear 'in the right way and as principle directs, for the sake of what is right and honourable' (ibid.: 128). Similarly, other virtues are also rational means between extreme and antisocial behaviours: 'modesty' falls between shyness and shamelessness, 'friendliness' between obsequiousness and cantanker-ousness, 'truthfulness' between understatement and boastfulness, and so on. In each case, living a life in accordance with these virtues helps the individual to maintain a proper balance between their own individual flourishing and the well-being of the political community. This need to maintain the social relations that make us human also explains why 'justice' is a particularly important virtue for Aristotle since it is that which determines who has an excess (more than they ought to have) or a deficiency (less than they ought to have) of the things necessary for their well-being in general.

For our current purposes, however, all we need to recognize is that Aristotle links virtues to ideas of human nature, individual character formation and social responsibility. Striving to live a virtuous life helps us to be happy and fulfilled and attain our full potential as a human being. It also ensures that we will act in a responsible and appropriate way towards others. This link between character formation and social responsibility is important because it relates individual and social moral values. It can also be understood as requiring what Michel Foucault termed a 'practice of the self' (1996: 432). By this Foucault meant that rather than simply following or applying moral rules like 'thou shalt not kill', the person concerned engages in a process of moral self-development and character formation over time. They choose to work on themself through self-reflection, sculpting their self-identity over the course of their life in order to make their character ever closer to an ideal model of what a virtuous person should be like. The exact form that this ethical work (*travail éthique*) on oneself takes varies quite widely. It may involve meditation, moral 'thought exercises', or even physical hardships, but the point is always gradually to transform yourself in accordance with a model of virtue. In this way you attain the desired characteristics that make you a virtuous person, someone who has fulfilled their proper human potential.

This idea of character formation survives in different forms up to the present day. Pierre Hadot traces its changes through the various schools of philosophy in ancient Greece, which he argues not only offered different ideals by which to live, but taught various 'spiritual exercises' that acted almost as a form of moral therapy to help individuals transform their existence. These exercises usually involved trying to learn how to control one's passions, one's 'unregulated desires and exaggerated fears' (Hadot, 1995: 83). Although each philosophical school 'had its own therapeutic method, all of them linked their therapeutics to a profound transformation of the individual's mode of seeing and being. The object of the spiritual exercises was to bring about this transformation' (ibid.).

For example, the Roman emperor Marcus Aurelius as a follower of the Stoic school of philosophy regarded self-control and fortitude in the face of adversity as important character traits, hence our contemporary use of the term 'stoical'. His famous *Meditations* provides a series of passages intended to encourage the reader to think about their (ultimately quite insignificant) place in the universe, and thus eventually distance themselves from avarice, pride and other character defects.

> These foods and dishes . . . are only dead fish, birds and pigs, this Falerian wine is a bit of grape-juice; this purple-edged toga is some sheep's hairs dipped in the blood of a shell-fish; as for sex, it is the rubbing together of pieces of gut, followed by the spasmodic secretion of a little bit of slime.
>
> (Aurelius in Hadot, 1995: 185)

Here everything that is usually thought valuable and luxurious, including the symbol of the emperor's own power, the purple toga, is reassessed and brought down to earth. The point of such pessimism was not to make one depressed but to distance oneself from that which is frivolous and to concentrate the mind on living a virtuous life.

In a very different context Hadot also argues that St Augustine's *Confessions*, written several hundred years later, can also be understood in this manner, not as an autobiographical account of his life, but as a model of self-formation through religious enlightenment. The *Confessions* is effectively an account of St Augustine's conversion that describes his passage 'from a formless state to a state of formation' (Hadot, 1995: 17).

In early modern times this notion of self-formation and moral transformation becomes closely associated with the idea of culture itself. In becoming a cultured or cultivated individual one engages in a project of self-improvement that is both educational and moral. Thus Anthony Ashley Cooper, the 3rd Earl of Shaftesbury (1671–1713), argued that 'by improving we may be sure to advance our worth and real self-interest' (Shaftesbury, 1999: 332). Shaftesbury too likened the process of moral improvement to the workmanship of artists or artisans. Just as they look beyond the raw materials that are presented to them, seeking to transform (improve) them into something else – a work of art or craftsmanship – so we should work upon morally improving ourselves. We must try to transcend and excel ourselves by cultivating the virtues within us through a process of self-reflection. What Foucault would term the 'ethical work' here was to take the form of an inner conversation leading to self-knowledge. This self-knowledge was 'not an end in itself; rather it helped the individual refashion the self on a moral pattern' (Klein in Shaftesbury, 1999: viii). In this way, the person who learnt to cultivate a moral sensibility became more cultured both in the sense of becoming more 'civilized' and in terms of having a better sense of the needs of other members of that person's culture. That is, they simultaneously became a better person and attained and acted according to what Shaftesbury referred to as a *sensus communis* – literally, a practical sense of how to relate appropriately (ethically) to other members of their community.

We can trace this relationship between the self-formation of an individual's moral character and culture through the Enlightenment thinkers of the eighteenth century and into early nineteenth-century Europe. As we might expect from Durkheim's and Bauman's arguments, as we move into the modern age individuals have to take more and more responsibility for their own ethical education, for the self-formation of a virtuous character. Hans-Georg Gadamer argues that this process is evidenced in the increasing and changing use of the German term *Bildung* over this period. *Bildung*, which originally meant 'formation' – as in the development of natural forms, like a well-formed body or a rock formation (*Gebirgsbildung*) – comes to be more closely associated with culture, indeed so closely that it eventually becomes synonymous with it. Thus in German one can speak either of *Kultur* or *Bildung* (culture as a formative or educational experience), and early nineteenth-century philosophers like Hegel (1770–1831) began to use terms like *Sichbilden* to refer to educating or 'cultivating' oneself. At about the same time, a whole literary genre known as the *Bildungsroman* – that is, novels dealing with 'one person's formative years or spiritual education' (*Oxford English Dictionary*) – also became very popular. The archetypal examples of the *Bildungsroman* were Goethe's *Wilhelm Meister's Apprenticeship* and its sequel, *Wilhelm Meister's Travels*. In fact, travel – the idea of widening one's horizons and investigating one's cultural history – plays a key role in many of these tales of self-formation where life is metaphorically regarded as a *journey* from immaturity to maturity, from ignorance to understanding.

The acquisition of culture – becoming a 'civilized' or educated person – becomes closely tied in modernity to the idea of travel, of seeing the world and thus better understanding one's place in it. In the late eighteenth and early nineteenth centuries the notion of the Grand Tour, as a kind of rite of passage to adulthood, became influential among the upper echelons of European society. The offspring of the wealthy in Britain would travel abroad, particularly to Italy and Greece, 'not only to see classical sights, but to learn languages, manners and accomplishments, riding, dancing, and other social graces' (Graburn, 1989: 29). The Grand Tour was not merely, or even primarily, something one did for pleasure; it was not a holiday in the contemporary meaning of the term but an exercise in character building (Dent, 1975). It was an educative experience with strong moral overtones intended to help produce an autonomous individual with a 'refined' aesthetic and moral sensibility. The (not inconsiderable) effort required to travel in those times can also be seen as a form of ethical labour necessary to bring the as yet formless individual into contact with the refining cultural influences of past and present centres of artistic and educative excellence like Paris, Venice, Rome and Athens.

This tradition spread from the aristocracy to the new middle classes that emerged in the nineteenth century. John Ruskin, perhaps the foremost art and social critic of Victorian England, wrote a number of works cataloguing the morally edifying effects of culture on the traveller. His *Stones of Venice* was a particularly influential early 'guidebook' to those travellers visiting north-western Italy and wishing to familiarize themselves with its architecture, sculpture, history, and so on. But it was just as much a moral as an aesthetic guide to the wonders of these cultures. Indeed, Ruskin even describes the history of Venice in terms of the city's moral and religious improvement and decline:

> On this collateral question I wish the reader's mind to be fixed. It will give double interest to every detail: nor will the interest be profitless; for the evidence I shall be able to deduce from the arts of Venice will be . . . that the decline of her political prosperity was exactly coincident with that of domestic and individual religion.
>
> (1888: 9)

In other words, the traveller was to understand that the aesthetic, economic and ethical 'improvement' of the (Venetian) individual and their society went hand in hand, and this lesson was something they too should reflect upon for the benefit of their own moral character.

Under the influence of the Romantic poets like Byron, Shelley, Wordsworth and Keats, this educative and moral experience also extended to nature. The high mountains in particular, like the Alps, were regarded as the sources of *sublime* experiences – that is, experiences that transcended the everyday and carried one outside of oneself, allowing one a vision of a larger world and one's place in it. These sublime experiences were transformative; they made a lasting impression on the traveller's aesthetic and ethical values. Here too Ruskin popularized the notion that developing an appreciation of natural scenery was a necessary feature of acquiring moral maturity, of being virtuous. Although

> the absence of a love of nature is not an assured condemnation, its presence is an invariable sign of goodness of heart and justness of moral *perception* . . . that in proportion to the degree in which it is felt, will probably be the degree in which all nobleness and beauty of character will also be felt . . . and that wherever the feeling exists, it acts for good on the character to which it belongs.
>
> (2000: 29)

The point of travelling was not simply to visit places and bring home souvenirs to show one's relatives. The many beautiful sketches that someone like Ruskin produced on his personal travels were not just records of the architecture or natural scenery, like picture postcards. They were an integral part of the aesthetic and moral learning process of travelling; they represented a kind of ethical labour that he thought helped people develop necessary virtues. Indeed, the fact that travel was becoming easier (and less of a financial burden), owing to a much improved infrastructure, most especially in the form of a widespread railway network, actually detracted from its value in character formation so far as Ruskin was concerned.

> In the olden days of travelling, now to return no more, . . . distance could not be vanquished without toil, but . . . that toil was rewarded . . . [by] hours of peaceful and thoughtful pleasure, for which the rush of the railway station is perhaps not always, or to all men, an equivalent.
>
> (Ruskin, 1888: 57)

In other words, the fact that the journey was becoming increasingly effortless meant that it was less likely to impinge on the traveller's imagination. In addition to this, the very pace of travel ensured that the direct and immediate contact with the scenery and culture one encountered on the journey was replaced with a much more superficial and instrumental relation. The railways thus altered the relationship between travellers and the cultural and natural spaces around them in such a way as to reduce the morally beneficial effects of the journey. In Wolfgang Schivelbusch's words, the railways seemed to destroy 'that in-between, or travel space, which it was [previously] possible to "savor" while using the slow, work-intensive . . . form of transport' (1986: 38). The spaces between the railways' points of departure and destination now seemed to provide 'only a useless spectacle' to be seen, but not properly felt or experienced, from the railway carriage window.

Of course, criticisms like Ruskin's also evidence what might be regarded as an incipient elitism in the whole notion of striving to become more virtuous, which can often mean striving to be more virtuous than others. From Aristotle's time onwards the process of character formation has usually been envisaged in terms of separating oneself from the everyday and the ordinary, of becoming better and more virtuous than one's fellows through individual effort and strength of will. Virtue theory often seems to call upon something of an 'aristocratic' notion of the self. It regards the possession of virtues as an individual achievement, a moral mastery of the self that is analogous to musical virtuosos' mastery of their instrument. In this sense, a genuinely virtuous character is something that few can achieve. What is more, as MacIntyre remarks, the specific virtues recognized by Aristotle were not those relevant or expected of everyone in Greek society but those characteristics associated with the ruling nobility, a nobility that Aristotle himself belonged to. This was equally true of the values held to be important by those sending their offspring on the Grand Tour; they too had a specific notion of 'refinement', one that was as much an expression of their wish to be regarded as different from 'common' people as it was an expression of individual moral excellence. Thus the coming of the railways and the associated advent of mass tourism is regarded by commentators like Ruskin as a threat to moral development in yet another way. The individual traveller with time (and cash) to spare now finds themself competing for space in front of the painting or view with those with no moral purpose in mind. The mass tourist is not interested in the formation of their individual moral character; they are not a seeker after virtue but a pleasure-seeker who just wants blindly to follow the crowd.

To some extent, of course, one can still see a similar, though perhaps less extreme, distinction being drawn today between the genuine traveller and the somewhat more morally suspect tourist (Mowforth and Munt, 1998; Munt, 1994b). Those university students who set off on their own mini-versions of the Grand Tour with their inter-rail passes, or the backpackers in Thailand or South America, like to think of themselves as rather different from (and superior to) the average package holiday-maker who goes to spend two weeks on the beach. This difference is characterized precisely in terms of their attitude towards their travelling. They speak of travel in terms of its educational possibilities, and the hardships of sleeping on the railway station floor or the flea-ridden hostel become part of the ethical labour necessary for travel to be genuinely character building (see Scheyvens, 2002). Elsrud (2001) suggests that individuals in backpacker communities manifest culturally and socially constructed narratives about risk and adventure. These are displayed and performed by the backpackers through the consumption of new foods, a place, clothing, experiences, and so on. Backpackers and other kinds of independent travellers construct and explain their identities through the use of exciting stories of and actual acts of risk and adventure. This narration of identity is assisted by the ways that independent travellers rely on word of mouth for information (see Duffy, 2002: 20–46; Elsrud, 2001; Murphy, 2001).

This is clear in the growing number of holidays that have an element of work in them. Every week newspapers advertise sponsored bike rides through the Atlas Mountains of Morocco or treks in the Himalayas, with all proceeds going to a specific (but usually Western) charity. These vacations directly appeal to individuals who want what they define as a more fulfilling experience where they can feel that they have contributed *something* to the local environment or helped to raise finance for a charity at home. This intersection between vacationing and volunteer work can be regarded as one example of how individuals in modern societies construct and affirm their identities (see Desforges, 2000; Munt, 1994a, b). The ways that such tourists (or travellers) present their stories of risk and novelty further promote their sense of self as especially adventuresome, confident and capable (Elsrud, 2001; Munt, 1994a, b). For example, Coral Cay Conservation provides an opportunity for conservation diving holidays where volunteers learn to scuba dive, and undertake surveys of fish populations with the aim of establishing fishing quotas and protected areas for Belize (interview with Jonathan Ridley, director of Coral Cay Conservation, Calabash Caye, 25 January 1998). Interviews with the volunteers made it clear that they regarded their role as a means of learning about themselves as well as assisting conservation in a developing country. In the process of explaining their choice of vacation they referred to the attractiveness of the very basic nature of their existence on Calabash Caye.

For example, one conservation volunteer stated that she chose to volunteer for Belize instead of a project in the Philippines because life was much rougher in Belize. She perceived the Philippines conservation project as more luxurious because there were employees to do laundry and cleaning, whereas Belize 'seemed more hard-core, more self-sufficient, so I thought I would go the whole hog' (interview with Coral Cay Conservation volunteer Isobel, Calabash Caye, 30 January 1998).

The idea of travel as a virtuous enterprise still survives in cases like mountaineering and hill walking, where the individual relationship to the natural environment is maintained: 'It is said that it is part of one's education [*Bildung*] to see the Alps' (Simmel, 1997: 219). But Simmel, one of the founders of modern sociological thought, was extremely sceptical about the educational and moral value of alpine sports:

> In the alpine clubs there is the idea that the surmounting of life-endangering difficulties is morally commendable, a triumph of spirit over the resistance of the material, and a consequence of moral strength: of courage, will-power and the summoning of all abilities for an ideal goal.
>
> (ibid.: 220)

Those who believed this also tended to romanticize 'the good old days of alpine travel' (ibid.: 219), the 'difficult routes, prehistoric food and hard beds' (ibid.), just as Ruskin had done before them.

Despite the genuine enjoyment to be found in mountain scenery and the real feeling of achievement in surmounting a difficult peak, Simmel believes that these feelings have little to do with morality. They are linked to the fulfilment of selfish desires rather than ethically formative experiences. The problem lies in 'the confusion of the egoistic enjoyment of alpine sports with educational and moral values' (Simmel, 1997: 220).

> By distancing their own spiritual and educative values from other sensual pleasures . . . these people employ one of those easy self-deceptions whereby their own culture, which would find egoism shocking, retains a subjectivity despite its lofty sentiments and seeks shamelessly to cloak its own pleasures with objective justifications.
>
> (ibid.)

Simmel also argues that these experiences are not necessarily immediate, individual and 'authentic' but are mediated through contemporary social conditions. The 'power of capitalism extends itself to ideas as well; it is capable of annexing such a distinguished concept as education as its own private property', reducing the experience to one of egoistic (that is, selfish) pleasure rather than moral profundity.

Thus in a sense Simmel's critique brings us full circle back to the problems of amoralism and moral subjectivism discussed in Chapter 1 and the question of moral relativism discussed at the beginning of this chapter. His critique of the moral argument for mountaineering as a character-forming activity put forward by the alpine clubs shows how difficult it is to distinguish between 'practices of the self' that involve 'ethical work' and straightforward egoism that involves doing what one wants to do. Like Simmel, amoralists might argue that these alpinists have fallen victim to a kind of self-deception that mistakes their own desires for ethical ideals. Similarly, moral subjectivists might point out that in so far as contemporary travellers can choose which culture to visit and which landscape to appreciate, their travel effectively embodies little more than a set of subjective evaluations. Lastly, as Simmel again points out, the values we might adopt as a member of an alpine club are dependent upon prevailing social circumstances, such as the existence of a capitalist society, and hence appear morally relative.

This goes to show that there are few if any conclusive arguments where morality is concerned, but that should not be a counsel for despair. In fact, one could interpret this example in other ways. Clearly, mountaineering as a dangerous sport is often very far from being selfish in the amoralist's terms. The kind of (ethical) labour required to be a *good* mountaineer might indeed produce some extremely egotistical individuals who are competitive, proud of their self-sufficiency, and so on. But those same people will often risk life and limb for their fellow climbers; they are not simply self-interested.

Contra the moral subjectivist, the mountaineers do not just make up their values; they adhere to a strict, if unspoken, 'code' of conduct. There is a very real *sensus communis* among them, a recognition of what kinds of action are appropriate for particular circumstances. This sense of what is right, and of the virtues required by mountaineers, is precisely what has led climbers like Joe Simpson (1997) to condemn the current commercialization of mountaineering on Mount Everest mentioned in Chapter 1. He argues that this commercialization has led to a fall in moral standards, to the failure to respect the aesthetic and ethical aspects of mountaineering. He recognizes that his mountaineering heroes like Shipton, Tilman, Herzog, Shackleton, Whymper and Mummery 'were all highly ambitious' (Simpson, 1997: 200), but at the same time they also embodied the kind of character traits (virtues) that still make them moral exemplars for our con-temporaries. In this sense, then, the virtues still have an important place: they characterize what it means to be a good mountaineer or a good traveller, to belong to a particular community that engages in certain morally regulated social practices. Here, then, we can see the power and promise of virtue ethics but also the problems for their survival in a fragmented and economically driven world that is almost 'after virtue'.

3 The greatest happiness is to travel?

Chapter 1 having argued that moral values are real and important *social* phenomena, not just matters of subjective personal taste, Chapter 2 proceeded to examine the implications of this discovery for travel and tourism. If ethical values are 'social' products, doesn't this mean that different societies will have different and perhaps even incommensurable values? What happens, then, when cultures meet, as they often do in the course of tourism developments? What if the values of host and guest communities clash? Is it possible to judge who is right and who is wrong about contested issues? Such questions are bound up with the notion of 'moral relativism'. But relativism offers no easy answers since the call to tolerate such ethical differences simply is not a practical or a moral option in many cases and may actually run counter to the most deeply held beliefs and values of those concerned.

A second issue arose in relation to the manner in which individuals come to associate themselves with specific ideas of a virtuous life. Such virtues seem, historically at least, to be dependent upon the individual having a certain specified social role to play and having to live up to certain social expectations associated with that role. In some societies these roles have been relatively fixed, and specific virtues can provide a stable, though perhaps somewhat limiting, sense of moral purpose for the individual concerned. However, in modern Western societies people usually adopt multiple social roles, and these roles can change frequently over time, making it far more difficult for the individual to appeal to fixed ideals of a virtuous life. This is certainly the case when we examine tourists. It is clear that there are multiple types of tourists, and each person engages in performing a variety of roles during the course of a vacation. Hence it is important not to categorize tourists as a single group in terms of their behaviours or in terms of the ways their vacation choices impact on host societies (Cohen, 1979; McMinn and Cater, 1998). As we saw, Durkheim suggested that in many ways individuality is itself the only common ground between people in modern societies. However, he also argued that since we all share this idea (and ideal) of individualism, this can actually provide the foundation of a new kind of social and moral order based on respect for each person's individuality. This is, then, quite literally, a case of making a virtue out of a necessity.

With the coming of modernity, people seemingly found themselves in a situation where they were encouraged to engage in a special kind of 'ethical work' upon themselves in order to develop as individuals, as beings who should strive to become increasingly autonomous as they reach maturity. We suggested that the educational experiences and even the hardships associated with travel could thus take on a moral aura. As indicated in Chapter 2, Elsrud suggests that backpackers clearly display culturally and socially constructed ideas of risk and adventure to narrate their identities (Desforges, 2000; Elsrud, 2001; Hampton, 1998). This performative aspect can also be seen in terms of narrating a moral identity, since many define themselves as independent travellers, concerned to demonstrate their awareness of moral and ethical dilemmas associated with modern travel (Duffy, 2002: 2–46). With the advent of the Grand Tour, travel became virtuous because it exhibited the person's independence from their original social ties. It exposed them to different cultures and different ideas, to a history and scenery that could inform a novel view of their place in the world. But the point of such travel was not, as is sometimes thought, to immerse oneself in a different culture, but to observe it with a sense of thoughtful detachment and to use these observations for one's personal moral betterment. As Rosalind Krauss remarks, Ruskin maintained a 'contemplative abstraction from the world' (Ruskin in Krauss, 1996: 5) on his travels in Europe. The fact that he could not always understand what was being said only added to the feeling that the events around him were *acted out* for his benefit. He compared the scenes he observed in foreign cities to witnessing a 'melodious opera' or a 'pantomime' – both of which, we should remember, frequently take the form of moral tales. Opera was quite a suitable metaphor for Ruskin to use since, then as now, it was an acquired taste, maintaining a cachet as a culturally and morally edifying spectacle despite, or perhaps because of, the fact that the actual words were incomprehensible to most of the audience. If other peoples and places could also be amusing, as Ruskin's use of the 'pantomime' metaphor suggests, then so much the better.

However, as Simmel points out, the problem is that an ethical attention to self-formation in a moral sense often seems indistinguishable from pure egoism; the virtue of individualism easily falls into an amoral self-absorption or even downright selfishness. As Munt argues, the new middle-class traveller might be better defined as the new 'egotourist' concerned only with self-development (Munt, 1994a, b). The concentration on individual virtues will not necessarily help us to resolve the problem of moral differences within and across cultures, either. What one person regards as a virtue – for example, faith, temperance or obedience – is often anathema to another, and nowhere are such differences more apparent or more important than in the issues surrounding tourism development. Here decisions are made on a daily basis that threaten to compromise those things that people hold most dear, including their deeply held beliefs about right and wrong. We must therefore address questions of moral *governance*, namely, who is entitled to make such ethically controversial decisions and on what grounds are such decisions deemed legitimate?

It is precisely this problem of moral governance that has puzzled generations of modern philosophers. Rather than simply take sides over substantive moral questions or particular value judgements, they have usually responded by developing theoretical frameworks that seek to provide a general account of the key features of *all* moral systems, together with an impartial method of decision making. Such theories are sometimes referred to as '*meta*-ethical' because they claim to go beyond, or find the basis for, the 'normative' ethical values that constitute our day-to-day moral values. The obvious value of such theories is that if they are genuinely not tied to any particular social context they may be universally applicable and capable of evaluating ethical controversies in an objective and non-partisan manner. One such theory is utilitarianism.

Utilitarianism

As we have seen, even the notoriously moralistic Ruskin recognized that travel was about more than moral edification: it was also about enjoyment. Whatever image we may have of them, even the most staid of Victorian travellers wanted, where possible, to be 'amused'. Today, of course, enjoyment is seen by most as the primary if not the only reason for going on holiday. Holidays are all about escaping the tiresome routine of the workplace in search of pure unadulterated and unconstrained pleasure. Whether it involves lying flat out on a beach in the Algarve or dancing energetically around the Ibiza club scene, hedonism is the name of the game. Of course, it is true that people do not all get pleasure from the same things, and tourists do not form a single group with a clear set of similar interests (see Cohen, 1979; McMinn and Cater, 1998; Wickens, 2002). Some seek solace in the Scottish mountains, some want to ski in the Alps, some like the crowds at the English seaside resort of Blackpool, and others want a beach to themselves. But whatever our differences, holidays surely exemplify the fact that whenever possible we all seek pleasure.

This simple insight into the human psyche provides the basis for utilitarianism. Where some might regard the individual's search for personal pleasure as the very antithesis of genuine moral concern (the term 'hedonist' often carries with it an implied criticism), the early founders of the utilitarian movement, like Jeremy Bentham (1748–1832) and James Mill (1773–1836), regarded it as morality's very foundation. Bentham argued:

> Nature has placed mankind under the governance of two sovereign masters, *pain and pleasure* . . . they govern us in all we do, in all we say, in all we think. . . . The *principle of utility* recognizes this subjection and assumes it for the foundation of that system.
>
> (1987: 65)

Everyone, no matter who they are (or what culture they belong to), wants to avoid those things that cause them pain and to seek out those which give them pleasure. The principle of utility quite simply states that we should therefore approve or disapprove of every action according to its 'tendency to promote or to oppose . . . happiness' (Bentham, 1987: 65). In other words, what makes people happy is good, what causes pain is bad. And since, Bentham argued, a community is nothing more than the sum of the individuals who compose it, actions should be judged good if they tend to increase the overall amount of happiness or decrease the total amount of pain in that community. In Bentham's own words, 'it is the greatest happiness of the greatest number that is the measure of right and wrong' (1948: 3).

Utilitarianism is especially relevant in our later discussions of the ethics of sport hunting and community-based tourism in Zimbabwe (Chapter 7). For Zimbabwe, tourism is underpinned by a utilitarian philosophy. This utilitarian standpoint informs related conservation policies that aim to balance the needs of the rural poor with wildlife conservation policies that are supposed to provide the greatest good to the greatest number of people (Duffy, 2000b: 9–19; Hulme and Murphree, 2001).

Bentham was originally trained as a lawyer and his primary interest always remained law reform. At the end of the eighteenth century the law was incredibly complex, and what was deemed permissible was largely decided on the basis of tradition and legal precedence. Bentham recognized that immense social changes were taking place, changes which legal and governmental institutions must reflect if they were to accommodate these new political circumstances and avoid the bloody revolutions that occurred in North America and France. Rather than looking to the past for moral guidance, utilitarianism was a progressive doctrine which would, he believed, provide a simple rational basis for a new, improved legal system and a suitable principle by which all government institutions might operate. The greatest happiness for the greatest number would be a 'principle that is recognized by all men, the same arrangement that would serve for the juris-prudence of any one country, would serve with little variation for that of any other' (Bentham, 1948: 25).

In other words, utilitarianism promised a universally applicable way of making decisions about right and wrong that was flexible enough to take account of people's different and constantly changing needs. It argued that governments and the law should act in the best interests of the community as a whole while preserving individual freedoms. The role of law and of government was not to tell individuals what they should find pleasurable – it could not possibly do that – but to order society in such a way that the population's actual pleasures could be maximized and pains minimized. The utilitarian would not take sides in ethical disputes about what kind of action is right or wrong *per se*. No action was wrong in itself; it was wrong only if its *consequences* were detrimental to the general well-being. Where, for example, some people might argue that bullfighting is morally wrong and should be banned, the utilitarian response would be to see whether the distress caused by bullfights outweighs the pleasure of those watching; only if it does so could it be regarded as wrong. (Interestingly, Bentham famously argued that animals too could feel pleasure and pain, and so in this case the distress caused to the bull itself should also be added into the equation.) This seems to offer a solution to the problem of moral governance we raised above, because those who employ the principle of utility to make decisions claim to do so only on 'objective' and 'rational' grounds and can therefore claim to be impartial.

We can see from this limited discussion that utilitarianism has a number of features that seem to make it appealing when it comes to deciding moral controversies:

1   Universality – everyone agrees that pain is bad and pleasure good, so there seems to be a common (hedonistic) ground on which to find agreement.
2   Rationality – it relies not on culturally relative and potentially controversial ideas of right or wrong but on a simple process of adding the pleasure and subtracting the pain caused by any activity, a kind of moral accountancy sometimes referred to as a 'hedonistic calculus'. Just like the accountant, the utilitarian presents us with the ethical equivalent of a 'balance sheet' where if pleasure outweighs pain we make moral 'profit' and the action is deemed good. Where pain outweighs pleasure there is a moral 'deficit' and the action is bad.
3   Impartiality – as we have seen, this idea of rationality is also important in terms of utilitarianism's claim to be neutral over substantive ethical issues. Since it does not appeal to any potentially controversial traditional or religious ideas, it can also be non-judgmental about individuals' personal predilections so long as they do not decrease the general level of happiness.
4   Versatility – rather than setting out fixed and unchangeable ideas of right and wrong, it can take people's changing tastes into account, which is of course why it appealed to reformers like Bentham. It is also very sensitive to particular social contexts because it recognizes that an action that might prove beneficial in one time and place might not in another. It is always the consequences that count.

The utility of utilitarianism

Both advocates and critics of utilitarianism obviously recognize that its key feature is its claim that happiness can provide 'a common currency of moral thought' (Williams, 1982: 99). Potentially, then, utilitarianism offers us a cross-cultural method for objectively resolving the ethical conflicts arising from tourist developments. All we need to do is to investigate and quantify the amount of happiness produced by different development strategies. We should then be able to come to a rational decision about the best course of action to take by ensuring that we maximize the benefits and minimize the problems associated with the development.

To give one example, let us look at the case of Sri Lanka, where tourism developments have clearly had important social impacts over recent years. According to Ratnapala (1999), the local populace object to many of these changes, especially in terms of:

1   the gradual erosion of the social cohesion provided by traditional family and village ties;
2   increasing commercialization, meaning that people are no longer happy just

to satisfy their basic needs but get caught in a cycle of consumerism, constantly seeking more;

3   the perception that values like honesty, simplicity, respect for one's elders, and so on are being replaced with alien cultural values;

4   the problems associated with alcohol and drug abuse, prostitution and gambling.

On the positive side, tourism obviously brings in a great deal of money and creates a considerable amount of employment (though often in an uneven and inequitable fashion). Not least, of course, the tourists themselves derive a great deal of pleasure from their holidays. The ideal solution would seem to be 'to find the common ground where these pleasures of the tourists and the minimum expectation of a standard of behaviour that does not damage the pattern of [Sri Lankan] life . . . meet' (Ratnapala, 1999: 144). Could utilitarianism actually help develop or support such a solution?

It certainly seems that utilitarianism could provide support for those who want to reform the way tourism in Sri Lanka has been organized (just as it was used by Bentham's supporters as an argument for social reform in nineteenth-century Britain). We can see this if we look at some of the problems Ratnapala identifies. He argues that 'at the time when tourist resorts were planned very little attention was given to [the rural] villages' (1999: 139) affected by such developments. The result was an accelerated decline of traditional social institutions and practices, including traditional crafts like woodcarving. People in and around tourist resorts became economically 'chained' (ibid.) to tourism as a source of revenue. Those

Plate 3.1 Tourists elephant watching in Minneriya National Park, Sri Lanka, 2001

Plate 3.2 Another view of tourists in Minneriya National Park

traditional social practices that did survive came to be treated as resources whose purpose was primarily to cater for the tourists' need for an 'authentic' holiday experience (see also Chapter 6). Traditional festivals, dancing, clothing, crafts, and so on became increasingly commodified, treated as things to be bought and sold by entrepreneurial middlemen involved in the tourist trade. Traditional craft-workers on very low incomes were 'forced to forgo quality for quantity in order to satisfy the entrepreneur involved in marketing his products' (ibid.: 136) and to compete with 'pseudo-artists' lacking traditional craft skills making poor imitations.

Ratnapala suggests that in order to overcome such difficulties Sri Lanka needs a coherent 'cultural policy'. This policy will include strengthening village institutions and educating tourists so that they 'know how to behave in a certain area and what is expected of [them] in return for the deriving of the maximum satisfaction by [them]' (ibid.: 139). Such a policy would also support local craft organizations, thereby ensuring 'an equitable distribution of the advantages received by tourism among a far greater number of residents [rather] than among a few businessmen' (ibid.: 138). It is easy to see how the case for such a cultural policy might be put in the form of utilitarian arguments: first, because Ratnapala claims that educating tourists can minimize distress to locals and maximize the pleasures that can be derived from the tourists' experience; second, because strengthening local institutions can help allocate tourism revenues in such a way as to promote the greatest happiness for the greatest number of residents.

According to Ratnapala, the key problem seems to have been that no one took the consequences of the tourism development for many of the poorest members of society into account. As often happens in tourism developments, those in a position of power simply assume that their model of economic development will have a positive outcome for everyone, without actually investigating the needs of all those concerned (Akama *et al.*, 1996; Dieke, 1994; Sindiga, 1999). Since utilitarianism is inclusive in the sense that it includes in its calculations the consequences for *all* those involved, it might be argued that it also tends to support the genuine participation of the public in decision making. Bentham certainly thought that the whole community should be recognized as the ultimate tribunal for judging the rightness or wrongness of any decisions affecting them (Bentham in Peardon, 1974: 141).

However, if we examine these issues in more detail, then it becomes less clear that utilitarianism would necessarily offer support to the kind of cultural policy Ratnapala envisages, whether in terms of educating tourists, strengthening the position of traditional village and craft institutions, or greater public participation in decision making. If we take the case of education first, it is certainly true that many tourists do get pleasure from finding out about local customs, values and beliefs, and that such knowledge might well help stop them offending against local sensibilities – for example, by learning not to wear inappropriate clothing. However, a lot of tourists travel simply for sun and sand; their pleasure comes from having nothing to worry about, from being waited on by others, eating and drinking their favourite foods, and so on. This is perhaps especially true of those tourists who visit mass tourist destinations on charter flights (Krippendorf, 1987; Pattullo, 1996; Pearce, 1995). Far from wanting to discover more about local culture and fit in around it, such tourists usually expect Western amenities, and staff who are 'alert and solicitous to their *wants* as well as to their needs' (Smith, 1989: 13). The last thing such tourists want is to be reminded of the social problems their holiday might be causing; for them ignorance is bliss.

It seems, then, that it is at least possible to argue that the more tourists there are, and consequently the more acute the locals' problems become, the more the balance of the hedonistic calculus actually shifts against the kind of cultural policy that promotes tourist education. Unfortunately, the greater happiness of the mass of tourists arriving might best be served by trying to suppress the moral qualms of traditionalists in the host community rather than educating the tourist. Nor can the utilitarian make any appeal to the tourists' 'better' nature. While some might regard learning about another culture as something 'good in itself' and condemn the ignorance, laziness and even debauchery of many sun-seeking tourists, a utilitarian can take no such moral stance. If people get more fun from sitting on the beach reading the latest trashy airport novel than they would from submersing themselves in local culture, then utilitarians must accept this, because they hold that pleasure is the *only* good. All other things being equal, there is no way to judge between the pleasure experienced by the well-informed and conscientious 'traveller' and the tourist who spends their entire fortnight happily drinking cocktails at the hotel bar oblivious to all that goes on around them. In Bentham's own phrase, '[q]uantity of pleasure being equal, pushpin [a popular game of his day] is as good as poetry' (Bentham in Mill, 1987a). This inability of utilitarianism to distinguish between what some might count as 'higher' and 'lower' pleasures led some of Bentham's contemporaries to label utilitarianism a philosophy fit for pigs rather than people.

Similar problems arise if we think about the issue of traditional village crafts. By its very nature, traditional craft production is small-scale and often unable to cope with the demands placed on it by mass tourism. As a result, industries spring up to meet the tourist market with imitation crafts. This is clear in the ways that Shona stone sculptures in Zimbabwe are readily copied and sold by street hawkers. The street traders use the original designs of more famous stone artists to make multiple copies and then sell their wares at a cheaper price to the passing tourist trade. If the tourists are unable to distinguish the genuine article from mass-produced imitations, and if mass production ensures that supply meets demand at a reasonable price through convenient retail outlets, then a utilitarian case could easily be made to favour the development of non-traditional modes of production (Bunn, 2000; Healy, 1994; Teague, 2000). Once again, utilitarianism seems to favour *quantity* over *quality*, an issue that led John Stuart Mill (1806–73), chief among the generation of utilitarians following Bentham, to try to develop a subtler version of utilitarianism. Mill argued that utilitarianism could and should distinguish different qualities of pleasure. 'It is quite compatible with the principle of utility to recognize the fact, that some *kinds* of pleasure are more desirable and more valuable than others' (Mill, 1987b: 279). These more valuable pleasures were, Mill argued, not those bodily pleasures more fitted to pigs but those associated with one's development as a fully cultured human being, namely the pleasures of the intellect, the imagination, and the moral sentiments. 'Human beings have faculties more elevated than the animal appetites, and when once made conscious of them, do not regard anything as happiness which does not include their gratification' (ibid.).

It could well be the case that many tourists would indeed discover that their visit was enriched through attaining knowledge of their hosts' traditions and practices. If so, Mill's argument could clearly support a policy of educating tourists and maintaining the 'authentic' artistic creations of traditional craftspeople. But on the other hand, Mill's comments could also be regarded as an expression of a kind of cultural elitism or downright snobbishness on the part of people who believe their own experiences and tastes to be superior to others'. Why, after all, should pleasures of the intellect be regarded as 'more elevated'? Why regard those who buy traditionally made crafts as necessarily having better 'taste' than those who return from their destination with mountains of mass-produced souvenirs? We will return to these questions in the next chapter, but such distinctions obviously call upon assumptions about the ethical and aesthetic qualities of activities that go well beyond the scope of a simple hedonistic calculus.

There are complex sociological issues at stake here because these kinds of educational or aesthetic distinctions are almost always tied up with the production and maintenance of cultural differences between social groups. For example, some visitors distinguish themselves by word, deed and destination as 'independent travellers', often disparaging the mass of 'tourists' who take package holidays

(Duffy, 2002: 31; Mowforth and Munt, 1998; Munt, 1994b). Pierre Bourdieu, a contemporary sociologist, puts this issue of 'distinction' quite succinctly:

> The denial of lower, coarse, vulgar, venal, servile – in a word natural enjoyment . . . implies an affirmation of the superiority of those who can be satisfied with the sublimated, refined, disinterested, gratuitous, distinguished pleasures forever closed to the profane. That is why art and cultural consumption [including, we might add, tourism] are predisposed, consciously and deliberately or not, to fulfil a social function of legitimating social differences.
>
> (1998: 7)

So long as goods are manufactured traditionally they are likely to remain relatively expensive and exclusive, and so the pleasure derived from them is limited to a cultural elite who can often, as Bourdieu suggests, use them as a means of distinguishing themselves from others they regard as *déclassé*.

This point leads us directly back to the issue of utilitarianism's 'democratic' credentials in terms of its support for the kind of greater public participation that many would argue must form a key part of a genuinely ethical policy on tourism development. While Mill's notion of different qualities of pleasure might support some aspects of the kinds of policies envisaged by Ratnapala, it also calls upon certain culturally specific (and hence not universally held) presuppositions that seem to contradict the basic tenets of Bentham's original work. Bentham's emphasis on quantity at least has the advantage of appearing less judgemental and more democratic because, all things being equal, any one person's pleasure counts the same as any other's, no matter how they might derive that pleasure. Superficially, then, at least, the basic utilitarian model seems to foster an inclusive form of social egalitarianism.

But this democratic egalitarianism has its limits. In certain circumstances the principle of the 'greatest happiness for the greatest number' can be perfectly compatible with leaving a minority of people in absolute misery. The utilitarian can justify a few people suffering so long as their consequent unhappiness is more than compensated for by a rise in the majority's pleasure. (Indeed, in some circumstances, where a few people gain a great deal of pleasure at the cost of only a marginal loss to others, it may even support substantial social inequalities.) The new five-star Sheraton Hotel in Addis Ababa, Ethiopia (Plate 3.3), undoubtedly provides a great deal of pleasure to its guests, and may also benefit employees through providing them with an opportunity to earn money. However, the stylish and expensive hotel has attracted criticism because it sits uncomfortably with widespread and obvious poverty in one of the world's poorest nations. While its owner has been praised for his sensitive treatment and rehoming of the shack-dwellers who used to live on the site, the Sheraton is still surrounded by poverty and poor housing. Keefe describes a case near Banaue in the Philippines where approximately forty Ibaloi families were displaced when their land was appropriated for a golf course development intended to attract tourists. These

Plate 3.3 The new five-star Sheraton Hotel in the centre of Addis Ababa, Ethiopia, 2002

relocated people were offered only substandard bamboo housing in recompense for their traditional wooden houses, and many lost their previous livelihoods growing and selling mangoes and rice for local markets.

> These families were once self-sufficient with surpluses to sell, but are now struggling to grow enough food for their own needs. Some of them have been forced to take casual work for a few pesos a day at the golf course which brought about their financial ruin.
>
> (1995: 16)

Obviously the Ibaloi have suffered great losses here. But utilitarian arguments would not necessarily oppose the development of the golf course or protect the Ibaloi. It could be that the misery of a mere forty families is more than outweighed by the gains of entrepreneurs, new employees, the Philippines' economy in general and, of course, the consequent pleasure of many golfers.

It is this emphasis on consequences that is utilitarianism's great strength and its greatest weakness. (It is also why it is referred to as a variety of consequentialism.) On the one hand, it means that utilitarianism avoids the kind of stultifying moralism that responds to the question of why a particular kind of behaviour is wrong with 'it just is' or 'it has always been so'. Utilitarians are able to give a clear rationale for their decision and do not have to follow hard and fast rules. For example, imagine a person who had been told never to tell lies suddenly being faced with a situation where being truthful might initiate a conflict in which many people would die. Here telling a (white) lie seems justified because it is clearly the lesser of two 'evils' and the only way to avoid disastrous consequences. Utilitarianism certainly seems to support the decision to lie here because as far as the utilitarian is concerned there is nothing wrong with lying *per se*; only the consequences of our actions matter, and in this particular case lying clearly produces more happiness than telling the truth. However, while utilitarianism will often support and express our ethical inclinations in such cases, there are, as the case of the Ibaloi shows, circumstances where its single-minded emphasis on consequences appears to result in some bizarre, unprincipled and even unjust decisions.

To give a couple of examples: William Godwin (a contemporary of Bentham and the father of Mary Shelley of *Frankenstein* fame) was also a proponent of a variety of consequentialism. He argued that given a choice between saving a famous author (Fénelon) or that author's valet from the flames of a burning building, justice demanded that 'that life ought to be preserved which will be most conducive to the general good' (Godwin, 1976: 169). This meant that one should choose to save the author even if the valet happened to be one's own brother or father, because one's own personal loss should be balanced against the loss of a future literary masterpiece that could bring pleasure to thousands. But is this really where most people would think their moral duty lay?

Or consider the case of a dancing bear that has been badly mistreated during training so that it can later provide quaint photo-opportunities to visiting tourists and an income to its handlers. Is this mistreatment really justifiable on the basis that the bear's pain might be outweighed by the pleasure given to an unwitting audience? If so, wouldn't it also be justifiable on utilitarian grounds to mistreat humans in a similar way? Isn't it wrong to mistreat animals and other humans just to provide pleasure?

Of course, utilitarians can always try to counter such arguments. For example, they could argue that if we factored in the unhappiness of those tourists who feel sorry for the bear because of what they know about its harsh training regime, then this would outweigh any potential pleasure that might otherwise be gained. Beating the bear might thus be immoral on utilitarian grounds after all. But such arguments miss the critic's point because they still suggest that the rightness or wrongness of an action (beating the bear) will often depend upon circumstances that have only an accidental connection with that action (such as what proportion of tourists are actually aware of how bears are trained). Would bear beating really stop being wrong if no one knew about it?

Similarly, whale and dolphin watching has been the subject of much debate (Hughes, 2001). Whale watching undoubtedly generates revenue for operators, and these financial gains then promote an economic rationale for conserving whales to ensure that the whale-watching business is sustainable. However, the noise from the engines of the whale-watching boats can also disturb the whales, and the operators can knowingly or unwittingly get closer to whales than environmental laws allow. For example, in Tadoussac in Canada (Plate 3.4), boats regularly get close to whales and follow the individuals for long periods of time. Many tourists would be unaware of the precise regulations governing whale watching.

Plate 3.4 Whale watching in Tadoussac, Canada, 2000

To summarize our arguments so far: utilitarianism developed as a system of ethical governance that explicitly aligned itself with a democratic and reformist political programme in order to moderate the tensions associated with the emergence of modern society, tensions that today are frequently exemplified in tourism developments. But utilitarianism is democratic and egalitarian only in the sense that it will usually support what the majority (the greatest number) want. The case of mass tourism thus proves especially difficult for utilitarianism. The sheer weight of numbers involved, together with the fact that each resort is expected to offer its clientele a relatively homogeneous holiday experience, means that mass tourism is especially effective in imposing its values and expectations upon local communities (Lanfant, 1995). For example, in Belize a recently developed hotel on Ambergris Caye was attempting to arrange deals with major international tour companies. However, the companies insisted that the newly built swimming pool was three inches too shallow to conform to European safety regulations. The pool had to conform to European regulations so that their clients, who might get hurt jumping into the pool, could not bring legal claims against them. The companies insisted that the pool was dug out and the floor lowered by three inches, at a considerable cost to the local hotel owners. The view of the hotel was that demands of this kind were common when it was dealing with tour companies based in the industrialized world, indicating that the companies had no real understanding of the complex set of constraints that tourism businesses faced in the developing world.[1] Mass tourism is universally recognized as the most ethically problematic and least sustainable development option, but Bentham's emphasis on the quantity of pleasure produced might even seem to make it the preferred option!

Ironically, the greater the number of potential visitors, the less ground utilitarianism has on which to defend alternative values and social practices. This again serves to highlight the differences between Bentham and Mill because, as Urry argues, mass tourism 'has been thoroughly based on popular pleasures, on an anti-elitism with little separation of art from social life; it has typically involved not contemplation but high levels of audience participation' (1997: 86). In other words, unless they are willing to contradict their basic hedonistic tenets, utilitarians would seem to have to agree that the millions of sun-seeking people flying off on package holidays simply cannot be (ethically) 'wrong'.

Act and rule utilitarianism

It seems, then, that despite the advantages claimed by utilitarianism in terms of offering a simple, rational and versatile framework for evaluating ethical decisions, its wholesale application can sometimes support quite unsavoury decisions, especially where minorities are concerned. But additional problems arise if we actually want to employ utilitarianism as a method of moral governance. Any organization trying to create an ethical policy for tourism development cannot be expected to look at every aspect of every single development in all of its possible guises and with all of its possible outcomes. It will want instead to provide much more general guidelines or rules governing how, when and where developments might be judged ethical.

This distinction between individual cases and general rules has been a source of considerable debate among philosophers interested in utilitarianism – a debate usually expressed in terms of 'act utilitarianism' versus 'rule utilitarianism' (Lyons, 1965; Smart and Williams, 1990). As the terms suggest, act utilitarianism judges the rightness or wrongness of a particular action in terms of the consequences of that action alone. By contrast, rule utilitarianism suggests that we judge actions in terms of whether or not they comply with rules based upon evaluating the consequences occurring if everyone in similar circumstances acted in a like manner. For example, imagine a tourist who is contemplating visiting a country ruled by an extremely oppressive regime. (See, for example, an account of a tour in Burma by Doherty (1995).) This hypothetical regime is able to hold on to power only because it uses tourist revenues to buy arms and thereby suppress the protests of its citizens. Despite this obvious link between tourism and oppression, there might be (exceptional) circumstances where the *act* of a single individual visiting this country could be justified on utilitarian grounds. For the sake of argument, let us say that this particular person will add almost nothing to tourism revenues and actually use their holiday experiences to critique the regime when they return. Unfortunately, this kind of act utilitarianism may be of little use to anyone interested in producing more general guidelines – that is, in moral governance. They need to look at the broader picture, and it is clear that if everyone decided to travel to this country, then the oppressive regime is more likely to survive. Rule utilitarians would thus argue that it is not ethical to travel, basing their calculations on what would happen if everyone were to follow the rules that they set out.

Rule utilitarianism has been thought to overcome some of the objections levelled against act utilitarianism. First, it allows us to make general statements about the morality of particular kinds of behaviour such as visiting countries with oppressive regimes, murdering people, and so on, which most people (and especially those interested in governing any particular aspect of society) want to be able to do. Strict act utilitarianism cannot make such statements because it is not such travel, or even murder itself, that is wrong; it is only ever a particular action judged in the light of its specific consequences. (Which, of course, leaves open the thorny philosophical question of exactly what counts as an action – see Lyons, 1965: 30–61.) Second, in doing this, rule utilitarianism also appears to insulate the utilitarian from having to defend apparently immoral acts that might, because of exceptional circumstances, actually increase utility. Third, in providing us with rules of thumb to guide our behaviour it absolves us from the necessity of having constantly to calculate the consequences of our actions. Fourth, these rules can be embodied in legal frameworks or institutional guidelines.

The only problem is that these rules can exist only by virtue of their ignoring certain aspects of the particular contexts in which the decision is made. In doing this they seem to undermine the whole point of utilitarianism, which was to avoid unthinking reliance on precedent and examine each case on its consequentialist merits. As the above example illustrates, because it requires a level of abstraction from particular situations, rule utilitarianism will almost certainly deem some individual actions wrong that would have actually increased utility, and some other actions right, that decrease it. This has led some to argue that rule utilitarianism cannot be justified on utilitarian grounds, or alternatively that reframing the rules with sufficient sensitivity to avoid such contradictory outcomes would require them to be so context specific that rule utilitarianism must effectively collapse back into act utilitarianism (Smart and Williams, 1990: 12).

Rule utilitarianism certainly embodies a kind of bureaucratic compromise that often ignores individual circumstances for the purposes of moral governance, for ease of administration and in order to attain a legalistic clarity about what constitutes right and wrong. It also means that the power of making moral decisions is shifted away from the individuals concerned to those deemed to be 'experts' capable of calculating the overall gains and losses from implementing different legal rules or social policies. Rule utilitarianism thus easily becomes associated with an Orwellian notion of a 'Big Brother' scenario where the individual's freedom to make choices counts for almost nothing (Lukes, 1995). Why should this be? How can the rule utilitarian justify the need for decisions to be made centrally when the individual act utilitarian has the same ends in mind in passing judgement on their actions, namely seeking to maximize *social* (and not just their own individual) utility? Only by arguing that those making the rules are in some way in a better position to see the whole social picture, perhaps by having more information, or being better able to take into account the full repercussions of any individual act upon the subsequent strategies of others (Harsanyi, 1982).

Some further problems with utilitarianism

If utilitarianism is to be of use in deciding between different strategies for tourism development, we obviously need to know what the likely outcomes of adopting those strategies will be in terms of social utility. It is imperative that the utilitarian should be able to evaluate and compare the *consequences* of committing particular acts or following certain rules. But this is where the problems really begin. The first difficulty is that the utilitarian asks us to make decisions on the basis of what consequences will follow from our actions. But as we all know, humans are not prescient. We cannot foretell future events with any degree of certainty, and the further into the future those consequences unfold, the more the uncertainties must grow and the greater the difficulty in predicting and appraising them. These problems of predictability are exacerbated the larger, and more complex, the social system involved. Nor, since the consequences of any action may carry on into an effectively infinite future, is there any obvious place at which to stop and calculate whether the decision was right or wrong. Do we calculate its effects a week after we made the decision, a year, ten years? In other words, despite its air of mathematical precision, utilitarianism is almost inevitably going to rely on somebody's 'best guess' at the outcomes and a relatively arbitrary decision as to when to make our calculations.

Similarly, we may find our evaluations of the morality of past decisions changing as time goes by since they will inevitably be made from constantly altering temporal and social perspectives. Such problems are certainly not unique to utilitarianism, but they are much more serious because the utilitarian is *entirely* reliant on such calculations to determine the rightness or wrongness of a given decision or strategy.

There is a more fundamental problem still when it comes to calculating outcomes. It is all very well to speak vaguely of maximizing happiness, but can we really measure happiness in such a way as to make credible comparisons between different individuals and across different strategies? Is the happiness experienced in winning the lottery comparable to that of falling head over heels in love, or is the pleasure derived from a concert in any sense comparable to that derived from relieving the pain of patients in a hospital ward? Even if there is some relation between such disparate experiences, can these be expressed in a common hedonistic currency and measured in 'units' of happiness in the same way that economic gains and losses can be measured in euros or dollars?

Bentham did not actually suggest any such units, but he certainly believed that such comparisons were possible and measurable. He went into some detail about the variables such calculations would need to include, such as the *intensity*, *duration*, *certainty* and *remoteness* of the pleasure concerned. Obviously, from a utilitarian perspective, the more intense and long-lasting the pleasure, the better. By taking into account the degree of certainty of actions or strategies having the desired effect and how long it might take them to have those effects (how remote they were), Bentham also hoped to overcome objections about the difficulties in predicting particular outcomes. The utilitarian would simply need to apply some kind of weighting so as to give preference to those strategies that seemed more certain of bringing about happiness and would bring it about sooner. There were other variables including an action's *fecundity* (the chance of its indirectly producing other, similar pleasures/pains), *purity* and *extent* (the number of persons it affected).

It is unclear whether, or how far, early utilitarians like Bentham actually wanted to turn morality into a branch of applied mathematics. He did admit that 'it is not to be expected that this process should be strictly pursued previously to every moral judgement'; it should always be 'kept in view' in order to make the utilitarian project as 'exact' (Bentham, 1987: 88) as possible. Such talk of exactness and maximizing utility certainly helped give their theory an air of efficiency and a quasi-scientific aura that appealed to their age. But as Ryan (1987: 31) points out, John Stuart Mill 'never thought of Bentham as trying to give moral evaluation the certainty of a mathematical calculus'. Talk of a hedonistic 'calculus' was to be taken not literally but figuratively. It was just meant to clarify utilitarianism's general assumptions. However, an economically minded follower of Bentham did argue that the need to mathematize utilitarian comparisons could 'no longer be shirked' (Edgeworth, 1881: 7). Influential economists like A.C. Pigou and Alfred Marshall also tried to develop what Sen (1999) refers to as a 'mental metrics (of happiness or desire)' based on utilitarianism. These economic varieties of utilitarianism have been extremely influential in debates about development and welfare economics, so it is worth looking at the relationship between utilitarianism and development theory in a little detail.

Development, economics and utilitarianism

As we have seen, utilitarianism has, since its inception, been allied to an ideal of progressive social development, at least in so far as 'progress' and 'development' can be equated with increased social utility. The key theoretical and practical questions for those interested in utilitarianism's contribution to evaluating tourism development are thus, how do we define and measure social utility, and should increasing social utility be the (sole) aim of development? Bentham's utilitarianism is uncompromising in arguing that increased social utility is the *only* ethically justifiable aim, and defines it quite simply in terms of the greatest happiness for the greatest number.

While Bentham's formulation clearly has some serious drawbacks, the fact that happiness cannot be directly measured need not be one of these if, as some economists have suggested, we 'broaden' the notion of happiness first to something like 'the satisfaction of desires' and then to the 'satisfaction of personal preferences'. While happiness or desires are almost impossibly difficult to quantify, or even rank on a uniform scale, economists are used to measuring personal preferences in terms of the economic choices people make. As Sen explains,

utility is often defined in modern economic analysis as some numerical representation of a person's observable choices. . . . The basic formula is this:

if a person would choose an alternative *x* over another *y* then and only then that person has more utility from *x* than *y*. . . . [I]n this framework it is not substantively different to affirm that a person has more utility from *x* than from *y* than to say she would choose *x* given the choice between the two.

(1999: 60–61)

There is now, as Sen and Williams (1982: 12) note, 'a well-established tradition in modern economics of defining utility entirely in terms of choice', the key assumption being that individuals make choices in order to maximize their own personal utility. Many economists go further still and argue that since more choices are open to those with more money, we should be able to measure something approaching 'utility' in terms of real incomes. We can also compare people's preferences in terms of how they spend that money, or would spend it under idealized conditions, on certain commodities. If these assumptions hold, then it is relatively easy to calculate the effects of different development strategies on real incomes and so make comparisons which our new-found alliance between economics and utilitarianism can declare ethically as well as economically sound. What is more, it should be possible to govern the resulting socio-economic system in such a way that it conforms to the utilitarian ideal of the greatest good for the greatest number. For the economist, this might equate with a form of what is termed Pareto optimality – that is, where incomes or goods are so distributed that there is no possible alternative state in which at least one person is better off and no one worse off.

This economic perspective has all the persuasive simplicity that is so attractive about utilitarianism, with the added advantage that its calculations avoid the difficulties normally attendant on inter-personal comparisons of individuals' utility, such as having to compare their potentially disparate experiences of pleasures or desires. Perhaps unsurprisingly, given the importance of economic arguments, this kind of real-income/commodity approach holds great sway in decisions over actual tourism developments. But this, as Sen (1999: 69) points out, is where difficulties begin: 'The limitations of treating real-income comparisons as putative utility comparisons are quite severe, partly because of the complete arbitrariness . . . of the assumption that the same commodity bundle must yield the same level of utility to different persons.' People differ in many ways: those who are ill or old or pregnant need more than healthy individuals; differing environ-mental situations make vastly different demands on people; different cultural norms mean that the satisfactions people gain from the same resources may vary considerably. These, and other, variations must make something as simplistic as income 'a limited guide to welfare and the quality of life' (ibid.: 71).

This is not to say that income is unimportant when it comes to making decisions about development. It obviously is not since people's survival often depends directly upon their level of income. Nor is it to say that economists are entirely wrong in focusing their attention on individual choice. After all, 'the core defi-nition of development in the Human Development Reports of UNDP [the United Nations Development Programme] is "the enlargement of people's choices"'

(Pieterse, 2001: 6). But Sen's work does begin to challenge a number of the assumptions behind the reducing of ethical considerations to an economic model, and we can extend such criticisms further. First, monetary terms are not good at capturing what is meant by 'utility' in the original sense of happiness or well-being. Second, this model implies a contentious form of welfarism – that is, it judges the moral goodness of a state of affairs only on the basis of maximizing social utility. The first is a criticism of economists' use of utilitarianism, the second a criticism of utilitarianism itself.

Contra the economical appropriation of utilitarianism, it may be a platitude but it is nonetheless true to state that 'money can't buy you (or act as a measure for) everything'. Above a certain level of basic subsistence, living a fulfilling and happy existence is only incidentally dependent upon income or one's ability to choose different 'commodities'. Rich people are not necessarily happier, and nor, ironically, are those who make all their decisions on the basis of maximizing their own utility. As Bertrand Russell (1930) famously argued, happiness is usually something that we find incidentally through working for some other external end, rather than an end that can be sought in itself.

Indeed, the expansion of consumer choice and the commodification of the life-world in modern society have given rise to their own worries and stresses. The shopping malls in which we make so many of our choices between commodities (a new suit or a holiday) are a source of fear to many (Davidson, 2001). At most, Langman (1994: 67) argues, the shopping culture provides 'no more than intermittent palliatives for underlying anxiety and appropriation of, if annihilation of, subjectivity'. It is one of modern life's great ironies that the socio-economic over-emphasis on *consumer* choice may actually undermine the identity and autonomy of the individual making those choices. As commodity choice becomes central to our existence, so our existence itself becomes commodified, nothing more than an expression of consumption. '[C]ultural and touristic consumption reproduce the very fragmentation and isolation they would alleviate' (Langman, 1994: 68). This is doubly ironic when we consider that utilitarianism and neoclassical economics both developed to accommodate and express the modern age's new-found individualism, and both have contributed to elevating this same autonomous individual to almost iconic status.

This leads us on to another point, namely the dependence of this economical version of utilitarianism upon the idea that one can reduce people's ethical concerns to personal preferences. Since we argued against this move in some detail in Chapter 1 we will not reprise it here, but it is important to recognize that utilitarianism in general, and this version of utilitarianism in particular, risks falling back into a variety of moral subjectivism or even amoralism. Think, for example, of the economists' assumption that people make choices so as to maximize their own utility – that is, that they are making decisions on their own behalf and in their own self-interest. This is obviously not the case in many instances of ethical behaviour, and it is difficult to see how ethics can provide any defence against the over-zealous application of market forces if it finds itself reduced to free-trade morality.

The key argument against utilitarianism in all its forms is not, then, the charge of moral subjectivism, nor that it cannot provide an adequate measure of utility, nor even that such measurements do not adequately take into account the differences between people. Its most problematic assumption is its inherent welfarism – that is, the belief that only social utility matters. It is this problem that leads to all the arguments covered earlier in the chapter about how utilitarianism can apparently justify the oppression of minorities, terrible personal injustices and a wide variety of unsavoury social outcomes when not countered by other ethical discourses. It is to one of these discourses that we now turn, that of human rights.

4   Rights and codes of practice

As Chapter 3 showed, one of the principal problems with utilitarianism is that the well-being of minorities can be threatened wherever or whenever their needs or desires clash with those that might be deemed to facilitate a greater good. This means, for example, that utilitarianism can easily be used to justify the displacement of individuals or small-scale indigenous communities by tourist developments when the changes offer significant benefits for a region or country as a whole, even if the consequences for the minority are very serious. This seems quite wrong, because the leisure pursuits of tourists surely should not take precedence over the livelihoods or even lives of people already occupying an area. Golf simply is not as important as grain (Ling, 1995). Ethics must have a role in protecting the vital needs of minorities and/or those without political clout and economic power. This is usually seen to be the role of legislation and codes based on the recognition of human rights, and it is to these we now turn. We begin by examining the political origins and philosophical underpinnings of rights discourses and then turn to the relationship between rights and development and their influence on codes of practice in the tourism industry.

Rights

> We hold these truths to be sacred and undeniable; that all men are created equal and independent, that from that equal creation they derive rights inherent and inalienable, among which are the preservation of life, and liberty, and the pursuit of happiness.
>
> (Jefferson in Cohen and Cohen, 1960: 204)

These lines, attributed to Thomas Jefferson, form part of the original (1776) draft for the American Declaration of Independence. Similar sentiments were soon to be expressed on the opposite side of the Atlantic following the French Revolution. Article 2 of the 1791 French Constituent Assembly's *Declaration of the Rights of Man and the Citizen* claimed that the 'aim of every political association is to preserve the natural and inalienable rights of man. These rights are those of liberty, property, security, and resistance to oppression' (Douzinas, 2000: 86). Rights and revolutions have frequently gone hand in hand because the claim to possess and attempt to exercise a right to such things as life and liberty poses a direct challenge to those exercising power over others. As Low and Gleeson (1999) argue, the 'history of human rights is a history of struggle enacted over the spaces and territories of the globe' (see also Tuck, 1977).

Although such struggles are by no means over, rights discourses are now part of the vernacular of mainstream international affairs and, following the Second World War, have become central to debates over the question of 'development' in general (Chandler, 2002; Rosenthal, 1999). In 1948 the United Nations adopted the Universal Declaration of Human Rights (UDHR). This declared in almost identical terms to its revolutionary predecessors that 'recognition of the inherent dignity and of the equal and inalienable rights of all members of the human family is the foundation of freedom, justice and peace in the world' (Brownlie, 1994: 21). This has since given rise to hundreds if not thousands of documents each applying the idea of human rights to different geographic areas, cultures and activities, not least among them tourism.

In 1985 the sixth general assembly of the World Tourism Organization (WTO) meeting in Sofia adopted the Tourism Bill of Rights and Tourist Code. This document focused primarily on 'the rights to rest and leisure and the freedom to travel' (Handszuh, 1998: 1) while rather underplaying the concomitant responsibilities of tourists and tour operators and the rights of host communities. It drew upon previous UDHR declarations such as article 13, which states that 'everyone has the right to freedom of movement' (Brownlie, 1994: 23) and to leave and return to their own country. Indeed, the call for such rights to travel has a long history. The philosopher Immanuel Kant (1724–1804), who ironically never left his native town of Königsberg, had also argued for a *Besuchsrecht*, a right to visit foreign lands, to enjoy their hospitality and to be offered 'temporary sojourn' (Kant in Arendt, 1989: 16). Contemporary documents, such as the WTO's *Global Code of Ethics for Tourism*, have continued this line of argument while also trying to rectify the original imbalances between the rights of hosts and guests. Article 2 of this code states that tourism activities 'should promote human rights and, more particularly, the individual rights of the most vulnerable groups, notably children, the elderly, the handicapped, ethnic minorities and indigenous peoples' (International Association of Convention and Visitor Bureaus (IACVB), 2000). One critical element of the debate about the ethics of tourism development is balancing the rights of the individual to travel with the rights of host communities (see Butler and Hinch, 1996). Leisure time is widely regarded as a real need and a right, an essential break from everyday life (Krippendorf, 1987; Rojek, 1995a). However, this right to leisure, and tourism in particular, has a direct impact on host societies.

The same ethical concerns about universal human rights that played a key role in the political development of the modern world continue to be regarded as important in many areas of life, including tourism development. However, the form in which rights have been articulated has changed markedly from the eighteenth century to today. The American and French declarations were both individualistic and egalitarian in intent. Rights were something that each and every individual possessed prior to and independent of any formal recognition of their existence by governments, law courts or international bodies. These were *natural* rights, rights one possessed simply by dint of being human. Such rights were thought part of each individual's natural legacy, their human nature. This is why the political philosopher John Locke (1632–1704) justified the existence of certain rights on the basis of their having supposedly existed in humanity's original 'state of nature', that situation prior to any social organization having evolved or developed. To

> understand Political Power right, and derive it from its Original, we must consider what State all Men are naturally in, and that is, a *State of perfect freedom* to order their Actions and dispose of their Possessions, and Persons as they think fit, within the bounds of the Law of Nature, without asking leave, or depending upon the Will of any other Man.
>
> (Locke, 1988: 269)

Locke claims that people are naturally independent, free to make their own decisions about what to do with themselves and their possessions. They are *possessive* individuals in the sense that they are the sole proprietor and private 'owner' of themselves and the products of their individual labour. Locke thus argues that in addition to the right to liberty and life there are also natural property rights that exist prior to their recognition by, or even prior to the existence of, social and legal systems. Ownership was natural in so far as things belonged to individuals who had 'mixed' their labour with them – that is, had worked upon them in order to make them useful. Indeed, Locke argues, it was people's wish to extend and secure these property rights that originally led them to gather together in a society governed by a 'social contract' formally setting out their 'rights' and responsibilities to each other.[1]

It was precisely this kind of argument that drew the criticism of Jeremy Bentham (see Chapter 3), who was extremely sceptical of such 'natural' rights and worried by some of their (revolutionary) social implications. Although in later life Bentham was often associated with radical causes, he was scathing about the declarations of rights emanating from post-revolutionary France. They were 'anarchical fallacies', both dangerous and wrong. The fact that such rights were deemed part of our *natural* heritage and therefore to exist prior to and independently of their recognition by any governments or laws meant that they allowed people to challenge the authority of *all* governments and legislators. If rights were natural, then they could neither be forfeited by the bearer of those rights (they were in technical terms *inalienable*) nor annulled by any other authority (they were *indefeasible*, or in Bentham's terms *imprescriptible*).

Yet Bentham thought that if there was to be genuine social progress, then the actual 'rights' recognized by the law would have to change over time, and such changes should be based only on principles dictated by social well-being – that is, by utility. Thus so far as Bentham was concerned, the choice was between, on the one hand, legislating for gradual social progress with the welfare and happiness of the general population in mind and, on the other, constant social upheaval caused by those who used the language of rights as an excuse for resisting governmental authority. 'What then was their [the writers of the French Declaration] object in declaring the existence of imprescriptible rights. . . . This and no other – to excite and keep up a spirit of resistance to all laws, a spirit of insurrection against all governments' (Bentham in Parekh, 1973: 270).

Bentham's attack on the idea of natural rights was simple and straight to the point. '[T]here are no such things as natural rights, no such things as rights anterior to the establishment of government, no such things as natural rights [as] opposed to . . . legal [rights]' (Bentham in Parekh, 1973: 268). He understood why some (oppressed) people might wish that such rights did exist, but that was simply a case of wishful thinking. Having 'a reason for wishing that a certain right were established, is not that right; want is not supply, hunger is not bread' (ibid.: 269). In so far as rights existed at all, Bentham argued, they were social creations, not the natural properties of people. They owed their existence to governments granting such rights to their citizens and therefore could be changed by those governments. Natural rights were, Bentham concluded, 'simple nonsense, natural and imprescriptible rights, rhetorical nonsense, nonsense upon stilts' (ibid.).

Generally speaking, even those adverse to Bentham's utilitarian approach have found the fundamentals of his critique of natural rights convincing. It is difficult to envisage how a right could be 'natural' in the sense that it could be present before any society exists to recognize it, especially if we are talking about rights to things that seem to be socially defined, like property. Most contemporary rights theorists have abandoned the idea of natural rights, regarding them instead as thoroughly social and political constructs. There are exceptions to this, most notably Alan Gewirth (1978), who still argues that certain features of our human nature lead us inexorably to recognize the rights that all humans have to freedom and well-being, but such ideas are, as Habermas (1993: 150) remarks, 'untypical and rather easily criticizable'. In Alasdair MacIntyre's words, the 'best reason for asserting so bluntly that there are no such [natural] rights is indeed precisely the same type of reason which . . . we possess for asserting that there are no unicorns' (MacIntyre, 1993: 69), namely, that there is no evidence whatsoever for their existence.

Today, the notions of human nature deployed by Enlightenment philosophers like Locke seem unable to provide a philosophical foundation for human rights that can possibly claim to be independent of historical and social circumstances. Locke's kind of 'human rights foundationalism' is now, as Argentinian jurist Eduardo Rabossi claims, both 'outmoded and irrelevant'. Richard Rorty (1999: 70) too denies the 'existence of morally relevant transcultural facts' and agrees with Rabossi that it is not even 'worth raising' the question of 'whether human beings *really* have the rights enumerated in the Helsinki Declaration' (ibid.: 69). Yet ironically, just as the philosophical foundations of natural rights seems to have faltered and fallen apart, we have entered a world where discourses, claims and charters of human rights proliferate and take on increasingly international importance. We now inhabit a global 'human rights culture' (Rabossi in Rorty, 1999).

What, then, does this mean for the status of these rights? Certainly those wielding institutional power in governments, the United Nations, trade organizations, and so on now seem happy to endorse a plethora of rights claims in ways that their historical counterparts were not. This may perhaps be optimistically regarded as a sign of general social and moral progress. Or more pessimistically it could be regarded as part of a wider project of global governance that seeks to extend Western notions of rights, progress, development and economic management, among others, through the use of international organizations and global legal regimes (Chandler, 2002; Duffield, 2001). Such a project is also discernible in the discourse on the rights to travel and tourism. The World Tourism Organization can be regarded as one example of global governance. The Tourism Bill of Rights and Tourist Code as well as the Global Code of Ethics for Tourism represent a global and universalist approach to the debate about rights and tourism. Furthermore, it could be argued that its discourse about rights and tourism reflects a set of norms and values that have arisen out of modernity and then been globalized through codes and bills that are deemed to have global applicability (see Handszuh, 1998; IACVB, 2000). Since such codes might be regarded as culturally embedded, they can fail to provide a platform for critiquing forms and impacts of tourism development or specific tourist behaviours. Such codes can constitute a highly contested narrative of the inter-relationships between tourism and ethics.

If rights are no longer deemed natural, inalienable or indefeasible but are, as Bentham suggested, simply the product of social institutions, then this leaves those institutions in very powerful positions as society's moral arbiters. The fact that rights come to be regarded as social phenomena lacking any universal socio-political legitimacy coincides with the loss of at least some of their radical potential. Rights can no longer pose fundamental ethical challenges to institutions if those same institutions are supposed to be the source of their more limited ethical authority. The need to administer such rights also requires and helps to generate a massive increase in legal/bureaucratic structures. And as Max Weber argued, 'as an instrument for "societalizing" relations of power, bureaucracy has been and is a power instrument of the first order – for the one who controls the bureaucratic apparatus' (1964: 228).

A sceptic might argue that despite its radical and revolutionary origins the discourse of rights has been co-opted by those in power, who are able to choose which rights to grant and how (or whether) to enforce them. This kind of scepticism reinforces our point that however universal the language in which they are situated, rights claims are often politically contested and far from impartial in their application. Eighteenth-century revolutionaries and contemporary governments alike have both deployed discourses of rights in order to claim the moral high ground – that is, to argue that their actions are guided by ethical principles that transcend self-interested political considerations. However, their (political) actions often speak louder than their (moral) words. The French revolutionaries were happy to proclaim the rights of man but repressed and even guillotined those women who claimed similar rights (Le Dœuff, 1990). Many American revolutionaries refused to extend their definition of man to include black slaves. And, in more recent times, countries like the United States and Britain have roundly condemned dictators in some parts of the world for human rights abuses while happily supporting regimes with equally notorious records as and when it suits them. For example, the United States and British governments certainly did not regard their role in overthrowing the democratic governments of countries like Chile or Guyana, or even the US-sponsored Indonesian forces' use of geno-cide in East Timor, as breaches of human rights (Galeano, 1973; O'Shaughnessy, 2000; Pilger, 1992).[2] The New Labour British government's much-vaunted 'ethical' foreign policy did not stop it from going ahead with a deal to sell Hawk jets to the Indonesian government in 1997 (see Abrahamsen, 2001, for further discussion).

This issue of moral governance intersects with broader discussions about the rights and duties of the private sector in tourism development. While we may con-centrate on the individual responsibility of tourists to engage in ethical behaviours, there is also a need to examine the role played by notions of corporate respon-sibility (see Carr, 1968; Ciulla, 1991; Duffy, 2002: 155–160; Stabler 1997). Increasingly, multinational corporations have been under pressure to draw up codes of practice and codes of ethics that demonstrate their commitment to culturally, socially and environmentally sensitive business practice (*The Economist*, 20 July 1996: 63–64). This issue of responsibility regularly arises in tourism. For example, as tourists disembark at the El Tatio Geyser in Chile (Plate 4.1), does responsibility for reducing the environmental impact of their visit rest with the individual visitor, with the tour guide, or the tour company that markets and organ-izes the trips to Chile? We return to this central issue of corporate responsibility later in this chapter.

Plate 4.1 Dawn at the El Tatio Geyser, Chile, as the tourists disembark, 2000

Difficulties in apportioning (and deliberate attempts to avoid) responsibility for upholding rights have led Douzinas to claim that

> the moral claim [of human rights] is either fraudulent or naïve. Experience tells otherwise: human rights, like arms sales, aid to the developing world and trade preference or sanctions, are tools of international politics used . . . to help friends and harm enemies. . . . The foreign policy of governments is interest-led and as alien to ethical considerations as the investment choices of multinational corporations.
>
> (2000: 128)

From Douzinas's perspective, rights are part and parcel of international *realpolitik* – that is, they are often deployed as a smokescreen to cover the ruthless and self-interested pursuit of policies aimed at furthering particular rather than universal human interests.

However, it is important to note that Douzinas is not a moral sceptic in the sense outlined in Chapter 1. He believes that ethical values play real and important social roles and accepts that human rights treaties are not entirely 'devoid of value' (2000: 144). His scepticism is directed towards the gap in rights discourses between political *rhetoric* and moral *reality*. The original power and promise of human rights seemed to lie in their universalism – the fact that their adoption should signify an acceptance that *all* people should be able to expect that certain basic needs will be met and certain responsibilities towards them will be fulfilled. Rights, whether natural or statutory, are supposed to provide safeguards and guarantees that apply equally to all humans. As Kant argued,

> There is nothing more sacred in the wide world than the rights of others. They are inviolable. Woe unto him who trespasses upon the rights of another and tramples it underfoot! His right should be his security; it should be stronger than any shield or fortress.
>
> (Kant in Norman, 1983: 122)

For Kant, rights were moral principles that expressed what he termed 'categorical imperatives' (unconditional or absolute commands) – that is, they were principles that could be extended to, freely accepted by, and command respect from, all rational people. So far as Kant was concerned, this was the very sign of morality in general. 'The supreme principle of the doctrine of morals is, therefore, act on a maxim which can also hold as a universal law. – Any maxim that does not so qualify is contrary to morals' (Kant, 1996: 18). In effect, Kant asks us to try a kind of thought experiment before adopting a moral principle, namely, would we be willing that that same principle should be applied consistently and without exception to everyone in all circumstances, including ourselves? For example, as a rational person I certainly don't want to be killed, but I should also recognize that this applies equally to all other rational people. I can't, then, without contradiction, expect people to recognize my moral right to life if I don't recognize a corresponding duty on my part to respect the lives of others. Morality and rights should be impartial and tolerate no exceptions; the very idea of a right that applies only to some people in some circumstances is therefore a contradiction in its own terms.

One has to respect Kant's attempt to be consistent in his derivation and application of moral standards. But while his arguments about the universality of rights should stand as a salutary reminder of the unethical and flagrant opportunism of those governments, institutions and individuals that try to hide their self-interests behind a language of morality, his position is clearly somewhat idealistic and overly formal. By this we mean that the end result of Kant's ethics is a set of hard and fast rights or rules of conduct with associated and inescapable duties that take no cognizance of our actual circumstances, or even (by complete contrast with utilitarianism) the consequences of our actions. One can imagine a Kantian park-keeper who, having issued a categorical imperative not to walk on the grass, then allows himself to be run over by an oncoming vehicle rather than step off the path. In doing so he certainly does his moral duty, but his dedication to the rules is hardly likely to get him much credit from those watching. As Wain (1995: 203) argues,

> Kant's formulation of ethics in terms of a categorical imperative which prescribes what our duty is and enjoins us always to act from the motive of doing our duty no matter what, dismissing all consequences and side-effects of that action or course of conduct as irrelevant, has generally been regarded as too narrow and too idealistic to lie within the possibility of ordinary human beings.

The problem with Kant's abstract formalism is not just that such rules/rights can place duties on us that are too onerous. Real problems arise from situations where different rights claims clash, where for example exercising a right to property might interfere with someone else's right to liberty (see Box 4.1, p. 82). Although one can imagine ways of arguing that we should prioritize some rights claims above others (for example, a right to life seems pretty fundamental), this still seems to entail compromising some people's rights in some circumstances. The formal mechanism Kant uses to derive rights is also problematic because, as many later philosophers have argued, if you think about wording it carefully enough, the categorical imperative can be given almost any content whatsoever. Our park-keeper could, for example, have saved himself a lot of trouble if he had worded his imperative 'I may step on the grass only when . . .' and then filled in a list of specific circumstances that could apply only to himself and/or oncoming vehicles. As Alasdair MacIntyre remarks, the possibility of this kind of reformulation means that 'in practice the test of the categorical imperative imposes restrictions only on those insufficiently equipped with ingenuity [to reword the imperative to suit themselves]. And this surely is scarcely what Kant intended' (1967: 198).

Rights and development

The philosophical criticisms rehearsed in the previous section do not necessarily undermine the importance of rights, or of trying, where possible, to achieve consistency in their application. However, they do bring home some of the difficulties inherent in trying to produce a set of rules/rights that will apply to everyone in all circumstances. This is of particular import for issues in tourism development, because some have argued that, far from expressing universal concerns, rights discourses are inherently biased. Western proponents of human rights often seem oblivious to the manner in which their discourse reflects their own cultural origins in modern capitalist societies. This criticism can take more or less radical forms. Karl Marx, for example, referred bluntly to talk of rights as 'obsolete verbal rubbish' (Marx in Waldron, 1987: 135). For Marx, the model of the 'possessive individual', which rights theories like Locke's and Kant's embodied, was not a neutral (universal) description of human nature. Rather, it was a particularly bourgeois conception of that isolated, autonomous and propertied individual so central to capitalist enterprise. Even the French 'Declaration of the Rights of Man' thoughtlessly recapitulated this same (far from revolutionary) conception of an 'egoistic man separated from his fellow men and the community' (Marx in Waldron, 1987: 147). We will return to some of the implications of such criticisms in Chapter 5 in our discussion of social justice.

Other theorists have taken a less radical line, wanting to retain the critical potential of human rights while recognizing that the kind of rights claimed by people will differ considerably according to their social circumstances. Donnelley (1999: 257) cites Adamantia Pollis's argument that 'First', 'Second' and 'Third World' countries could all be regarded as having different rights agendas. The 'developed' First World emphasized civil and political rights, together with the right to private property, the (at the time) largely socialist Second World emphasized economic and social rights, and the much poorer Third World emphasized self-determination and economic development. The latter two kinds of rights claims, Pollis argued, were primarily group oriented rather than individualistically based. This compromise position seems to have some attractions, since it recognizes a certain degree of cultural relativism in allowing conceptions of human rights to be developed locally. But of course this threatens to throw us straight back into debates about moral relativism (see Chapter 2), because recognizing culturally variable rights might actually undermine the rationale for declarations of universal human rights in the first place. Donnelley himself is scathing of such relativism and regards Pollis's three worlds approach as 'largely a misguided capitulation to ideologically motivated arguments that sought to use the language of human rights to justify oppression' (1999: 258). In other words, he thinks that apologists for socialist regimes used arguments about cultural relativism to deprive some of those populations of their full (individual) human rights, rights these populations later demanded for themselves with the fall of communist regimes around the world.

There are, then, at least two aspects to this question of deploying arguments about human rights in a development context. Historically, colonial Western

Box 4.1 Tourism development, outdoor pursuits and rights of access

Some of the difficulties arising from the clash of contesting rights claims are clearly illustrated in the case of the rapidly increasing demands for tourist access to public and private lands in order to pursue outdoor activities like hill climbing, mountaineering, and so on. Many countries, like Norway and Sweden, have guaranteed rights of public access that can be traced back to the Middle Ages, although the rationale for recognizing such 'rights' has changed markedly. It was only following the growth of the outdoor movement in the late nineteenth and early twentieth centuries that this *allemansrätt* 'became an important part of mass recreation' (Kaltenborn *et al.*, 2001: 419). Norway is unique in recognizing the right of public access in a specific piece of legislation, the Open-Air Recreation Act of 1957. As Kaltenborn *et al.* argue, this kind of freedom of access was sustainable partly because of low local population densities, but the advent of global tourism and the increasing commodification of landscapes for activities like downhill skiing, rafting and glacier walks all threaten this delicate balance. The debate has also become one of 'the rights of private property versus the rights of the public to engage in outdoor recreation' (ibid.: 429). Some sporting entrepreneurs want to charge people to enter certain areas but are prevented from doing so because of the presumed right of access. They claim that their right to make a living is being challenged. Ironically, this tension is also exacerbated by the growth of recreational homes in areas of scenic beauty. Those who buy such homes because of their recreational potential often then want to restrict access so as to maintain their private use of such pleasures. Such tensions have been especially marked along the coastline, where legal cases have so far tended to favour the right of access over that of privacy.

Such issues are complicated yet further by specific historical precedents and economic and cultural changes. For example, MacIntyre *et al.* (2001) describe the situation in New Zealand, where the growth of outdoor pursuits has coincided with both the neoliberal restructuring of the New Zealand economy and the emergence of an indigenous landrights movement. Tourism is New Zealand's main source of foreign exchange, with some 1.8 million international arrivals annually (February 2000 to February 2001), meaning rapidly increasing demand for leisure access. Meanwhile, the country's economic restructuring has been extremely influential in decreasing governmental regulation of land use and weakening already fragmented access legislation in favour of private landowners. The scene is further complicated by the fact that the contemporary Maori have launched claims for large sections of public lands as settlement for land originally confiscated from their ancestors following the signing of the Treaty of Waitangi of 1840. This treaty was supposed to guarantee the Maori a right to self-determination (*tino rangatiratanga*). One such claim, the Ngai Tahu claim, covers much of New Zealand's South Island, and, where successful, such claims could mean that 'Maori values will be incorporated into the management of the land' (ibid.: 445). For example, Ngai Tahu requested that mountaineers should not step on the summit of Aoraki (Mount Cook) because this was deemed disrespectful to a mountain traditionally regarded as an ancestor. Here, then, we have some very complicated competing rights claims: to economic freedom, traditional land rights and open access for leisure. The question of how such claims are to be resolved is extremely difficult.

powers certainly refused to recognize that indigenous peoples should have the kind of guarantees to life and liberty that they granted (at least nominally) to their own citizens. Over many centuries these peoples were murdered, raped, tortured, enslaved and exhibited as curiosities on the spurious grounds that they were not fully human and therefore not worthy of rights. They were referred to as 'savages', 'child-like', 'sub-human' or 'beasts', and widely regarded as being incapable of fully rational thought (Jahoda, 1999), all of which made it easier to exclude them from the dominant moral sphere. Given this unpardonable history, it may seem progressive and long overdue that non-Western peoples should now have their rights recognized as equals in our 'post-colonial' world by global organizations like the United Nations.

However, given its Western origins, it is possible to see this expansion of a doctrine of human rights as yet another form of cultural colonialism justified, as colonialism itself often was, as a part of a supposedly progressive 'civilizing' process. This brings us back into key debates about the very notion of 'develop-ment' (see the Introduction and Chapter 3). Some have argued that the whole notion of 'development' is an invention of the rich minority of largely Western nations used to force the poor majority to follow a pre-given and prescribed trajectory of economic growth and modernization that is often unsuited to their actual needs. In Latouche's (1993: 160) words, 'development has been and still is the *Westernization of the world*'. Escobar (in Peet, 1997: 76) argues that

> development can be described as an apparatus ... that links forms of knowledge about the Third World with the deployment of forms of power and intervention resulting in the mapping and production of Third World societies. ... By means of this discourse, individuals, governments and communities are seen as 'under-developed' (or placed under conditions in which they tend to see themselves as such) and are treated accordingly.

(See also Escobar, 1995; Lehmann, 1997.) The clear message in labelling a country 'under-developed' is that it needs to compare itself with and conform to the model of other, more 'developed' nations. But from the kind of post-developmental perspective employed by Escobar, the current emphasis on human rights might be regarded as just one more example of 'colonialist' interventionism whereby the West regards its own moral ideals as the sole paradigm to which all other morally 'under-developed' societies should conform. In other words, from this perspective the attempt to globalize a doctrine of human rights carries within it an implicit belief in the West's own moral superiority and actually serves to marginalize alternative moral traditions in the Third World (see Chandler, 2002).

These are complex issues. The idea of 'development', like that of global human rights, came to institutional prominence following the Second World War. Development aid and assistance was often justified in terms of promoting human rights, and rights discourses were similarly employed to justify the need for development. It is thus tempting to see 'rights' and 'development' as no more than mutually supportive rhetorics used by institutions like the World Bank to impose their own (Western) models of economic growth on the unsuspecting poor (Williamson, 1993). But does this mean that we should simply reject the Western models of development and rights completely, as post-developmental and post-modern perspectives might suggest?

There are certainly problems with this strategy. Even if we regard traditional developmentalism with (justifiable) suspicion, it is problematic simply to argue that 'the real effect of modern "development" policies on the poor has been a substantive increase in their suffering' (Spretnak, 1999: 91). Things are seldom so clear-cut. It is certainly true that Western interventions have frequently encouraged unsuitable forms of 'maldevelopment' (Shiva, 1989) which have exacerbated rather than alleviated poverty. But this does not mean that Third World poverty is not a real problem or that one should adopt an anti- or post-developmental position rather than explore alternative developmental models. As Pieterse (2001: 101) argues, while it may be true *up to a point*, as Shiva and others suggest, that it is 'the economism of development that is truly pauperizing', we should not let poverty alleviation itself 'slip off the map'. The danger with rejecting a developmental model out of hand is that

> we homogenize and romanticize poverty, and equate poverty with purity (and the indigenous and the local with the original and authentic). The step . . . to a moral universe is worth taking, but a moral universe also involves action, and which action follows?

The real question is whether discourses of development and of rights can actually bring about actions that help the world's poor and oppressed, always remembering, as critics like Shiva remind us, that this is not just a matter of economics, but of achieving sustainable livelihoods. To some extent the jury is still out on this question. Past developmental strategies, even those supported by global organizations like the World Bank, have had significant failings in these terms (Rich, 1993), and doubts persist about the status and efficacy of international human rights. But while discourses about rights and development have been closely intertwined, this does not mean that they have always been mutually sup-portive. Even those sceptical of the philosophical basis of rights, and the politics of their deployment, must recognize their value in the sense that they at least provide a way of expressing opposition to flawed development projects. Thus, for example, the Maasai, who were excluded from some of their traditional lands in East Africa by successive wildlife conservation policies, are able to couch their claim to 50 per cent of tourism revenues from these national parks in terms of their 'rights' (Yale, 1997; see also Brockington, 2002; Sindiga, 1999). So although there are different interpretations of what rights are, and what rights we might have, they do provide a language for claims making that has widespread recognition precisely because of their historical and cultural associations with now globally dominant powers. This language of rights has become increasingly important in recent years with the development of 'sustainable' and 'ethical' tourism (see Chapter 7) and with controversies over tourism in countries with despotic regimes (see Box 4.2). Burns (1999) argues that developing countries such as Eritrea have to decide whether to concentrate on 'tourism first' as a path to sustainable development, or examine tourism as one possibility within a broader framework of 'development first'. Furthermore, in any kind of tourism development, Burns suggests that the community should have ownership of the planning process, because without it the right of individuals to travel can result in communities having to suffer the consequences of inappropriate tourism (ibid.: 346).

Codes of ethics

While the philosophical status of rights may be in doubt, the language of rights certainly seems to have achieved international institutional recognition and is explicitly and widely employed in codes of ethics for the tourism industry. The WTO makes extensive references to rights in its Global Code of Ethics for Tourism and in its own statutes. For example, Article 3 of the WTO's statutes defines its role as 'promoting and developing tourism with a view to contributing to economic development, international understanding, peace, prosperity and universal respect for, and observance of, human rights and fundamental freedoms for all without distinction as to race, sex, language or religion' (IACVB, 2000: 2). The preamble to the WTO Global Code is littered with declarations of rights, and two of the code's ten articles are specifically entitled 'The right to tourism' (Article 6) and 'The rights of the workers and entrepreneurs in the tourism industry' (Article 9).

Malloy and Fennell (1998) carried out a content analysis of forty separate codes of ethics (a total of 414 individual guidelines) in the tourism industry and found that 77 per cent of all guidelines were deontological – that is, rights or principle based. The frequent reference to rights in such documents is a sure sign of their global currency since the wording of such documents is always a matter of sensitive negotiation and extensive compromise between competing interests. The WTO itself claims that its code has to be 'clearly defined to allow comprehension and assessment by ordinary citizens concerned about tourism development' and 'use terms and terminology to be understood to various cultures' (Handszuh,

Box 4.2 Burma (Myanmar): tourism and human rights

Burma (renamed Myanmar) is a good example of a country that has recently sought to use its 'exotic' associations, in terms of its incredible Buddhist temples, its ethnic diversity and natural beauty, and its romantic associations – the Road to Mandalay – to encourage mass tourism. In order to gain much-needed foreign revenue, the government declared 1996 'Visit Myanmar Year', and massive infrastructural development projects were put into place in an attempt to encourage up to 500,000 visitors. However, Burma has been run since 1988 by a military dictatorship that referred to itself as the State Law and Order Restoration Council (SLORC). Because of a lack of capital, the developments were often accomplished through unpaid forced labour and the relocation of entire villages, and included what many outside observers, including Amnesty International and journalists like John Pilger, documented as sustained human rights abuses. More than 200,000 people were moved from Rangoon against their will to free up valuable inner-city sites for building, and 120,000 people were conscripted to work on the Ye-Tavoy railway, building 110 miles (*c.* 180 km) of railway embankments. Between 200 and 300 of these people are estimated to have died during construction. Twenty thousand people, some as young as 13 and many of them chained together, were forced to clean out the moat around the city of Mandalay itself (Tourism Concern, 1995: 7). Aung San Suu Kyi, the democratically elected, but until recently confined, head of Burmese state and winner of the 1991 Nobel Peace Prize, called on tourists to boycott Burma and on governments to impose sanctions against it until democracy was restored (Pilger, 1998: 155). A sustained campaign by organizations like Tourism Concern meant that the treatment of the Burmese population was given widespread publicity, and Visit Myanmar Year 'was an outstanding flop', with several British tour operators pulling 'out of Burma for ethical reasons' (Wheat, 1998).

However, even in such a well-documented case of human rights abuse some people differed about whether travel to and investment in Burma was 'unethical'. Dr Naw Angelene, SLORC's Director of Tourism, claimed that Visit Myanmar Year would bring enormous benefits to the Burmese people. 'Roads will be wider, lights will be brighter, tours will be cleaner, grass will be greener and, with more job opportunities, people will be happier' (in Pilger, 1998: 182). James B. Sherwood, the chair of Oriental Express Group, at that time the principal British tour operator in Burma organizing river cruises on the Irrawaddy, also argued that 'those who travel bring their money to help the economy and the standard of living of the people' (ibid.: 187). Despite protests, Lonely Planet guides continued to publish their tourist guide to Burma on the grounds that it was up to individuals to make their own ethical decisions about whether or not to visit.

1998). Its stated purpose is 'to serve as a frame of reference for the stakeholders in world tourism at the dawn of the next century and millennium' (IACVB, 2000: 1). In other words, the WTO regards the document as expressing culturally universal ideals that, if not timeless, can at least set the ground rules in terms of the next hundred years or more. The moral absolutism so characteristic of the language of human rights thus sits well with the high moral tone of the document. It also, of course, allows the WTO to make explicit links with, and ask for endorsement from, organizations like the United Nations, which already have a role as the global moral arbiters of human rights.

However, it is always important to set such documents in a wider context. As we have seen, once one discards the notion of *natural* rights, human rights themselves become subject to social and institutional negotiation. In other words, the absolute protection that Kant argued rights should afford often becomes more apparent than real; they become open to differing interpretations and constant revision. This can be seen both in terms of responses to specific rights abuses and in rights-based documents like the WTO code. As the example of Burma (Myanmar) (Box 4.2) shows, even the most flagrant rights abuses can be subject to different interpretations and responses. One can deny or ignore their existence, deny that what is happening actually constitutes a rights abuse in that social context, admit they exist but refuse to legislate or 'take sides', or even claim that by carrying on as normal one is involved in a form of constructive engagement (as many companies/governments did for years in apartheid South Africa).

In an institutional context, explicit support for, and deployment of, human rights claims can give a document an aura of impartiality and perpetuity, a moral weightiness that it might otherwise be seen as lacking. But here too the door is often left open for revision, compromise and interpretation. Thus, alongside the millennial rhetoric, the Global Code also envisages the 'creation of a *flexible* follow-up and evaluation mechanism with a view to ensuring the *constant adjustment* of the Code to the developments of world tourism and, more broadly, to the *changing conditions* of international relations' (IACVB, 2000: 9; our emphasis). Its guiding principles are also less obligatory than an emphasis on rights and duties might suggest, since its implementation involves the 'making available to States and other stakeholders in tourism development of a *conciliation* mechanism to which they may have recourse *by consensus* on a *voluntary* basis' (ibid.: 9; our emphasis). The WTO set up a World Committee on Tourism Ethics to monitor and implement the Global Code and act as a 'watchdog' (IACVB, 2000: 10), but its provisions only state that members of the WTO *may* declare that they accept the committee's ruling as binding.

In other words, it is important to recognize the limitations of such industry-led codes when these are compared to the kind of protection that someone like Kant thought rights should offer. In general, despite the fact that they are drafted in a legalistic language, they are often voluntary agreements that are little more than aspirational documents serving as indicators of best practice. They often have limited penalties for non-compliance and are seldom legally binding in the way that, for example, the US constitution guarantees its citizens certain legally enforceable rights and freedoms through the courts. Indeed, the point of industries like tourism adopting ethical codes has largely been to demonstrate a commitment to social responsibility and an ability to be self-regulating, thereby avoiding the imposition of external legislation. This is one of the reasons why codes of ethics suddenly became such a hot issue in the business community in the 1980s following the industrial deregulation characteristic of the New Right under Reagan and Thatcher.

> Deregulation is often justified by the assumption that the market will discipline business and . . . industries can regulate themselves. If they didn't regulate themselves then new laws would be created. Prudent firms began to think about their operations from a moral point of view.
>
> (Ciulla, 1991: 72)

By 1986 75 per cent of the top 500 US companies were 'prudent' enough to have drafted a code of ethics.

This does not mean that such codes are worthless. As Ciulla (1991: 73) notes, 'On their own, codes don't do much and are sometimes ignored. However they can serve an important function if they introduce constructive dialogue about ethical issues into [organizations].' They also provide a tangible set of standards and principles that can be appealed to in order to justify or criticize a particular activity and can act as a kite mark for consumers (in so far as those businesses/ organizations that have signed up to these codes of practice can be assumed to comply with them). This is precisely why those tourism companies/destinations that seek to associate themselves with ethical tourism, ecotourism or sustainable tourism (see Chapter 7) are most likely to promote explicit ethical codes. This helps them distinguish their approach from that of the tourist industry at large by emphasizing their role in, for example, sustaining cultural and biological diversity, as can be seen in the Belize Eco-Tourism Association Code of Ethics (BETA, 1999; see also Forsyth, 1995; Wight, 1994). Doubts do, however, persist, not just about the ability of an industry like tourism to regulate itself, but also about a perceived incompatibility between 'being ethical' and 'doing business'. To some, 'business ethics' seems like a contradiction on terms, especially when one considers scandals like the recent Enron debacle in the United States.

Such doubts are not expressed just by those who regard business itself as morally dubious on the basis that 'profits' are only ever made at other people's expense. Some avid proponents of cut-throat capitalism also think that ethics has no place in business, or at least that 'the ethics of business are not those of society but rather those of the poker game' (Carr, 1968). In poker, bluffing is an acceptable gaming strategy often necessary for success, and, Carr argues, whatever the moral façade adopted by many businesses, they are actually run with only one thing in mind: success. The 'major tests of every move in business, as in all games of strategy, are legality and profit' (ibid.: 149); whatever isn't actually forbidden by the law is permissible. 'Violations of the ethical ideals of society are common in business, but they are not necessarily violations of business principles' (ibid.: 148). From this perspective the adoption of codes of ethics might be seen as a bluff on the part of business, a cynical public relations exercise or a self-serving strategy to 'win friends and influence people'.

This scepticism about business ethics explains why management theorists like Peter Drucker have dismissed it as a passing fad, or a form of 'ethical chic' (Drucker in Ciulla, 1991: 67; Sasseen, 1993: 31), and why, as Page and Dowling (2002: 243) note, 'some critics have argued that this may just be another marketing ploy for the tourism industry'. It is certainly true that ethics is being 'sold' within industry as a way of avoiding legislation and maximizing profits in a climate of increasing environmental and social concerns:

> The arguments used to win their [businesses'] support have little to do with philanthropy and everything to do with the bottom line. 'Our rationale is very clear,' says Robert Davies [in his role as chief executive of the Prince of Wales Business Leaders Forum]. . . . 'We believe a business case exists for getting involved in local communities. Multinational companies will be more successful globally if they act like good citizens locally.'
>
> (Sasseen, 1993: 30)

For many, business ethics is primarily about reacting to customers' values and expectations. Sasseen quotes the head of public policy at the National Westminster Bank, who argues that companies adopt codes of ethics because of 'the importance of social issues to a company's *image*, and the glare of publicity when problems arise' (1993: 31; our emphasis). Put simply, codes of ethics improve an industry's image, allow it to avoid scandal, and improve its sales.

Ironically, then, the need for industries like tourism to follow moral codes is often justified to the business community on economic, not ethical, grounds (see Chapter 1). This may be pragmatic, reflecting the fact that the 'bottom line' in business is money, but it is also problematic because it suggests very strongly that ethical principles are subservient to business principles. Everyone can be ethical when it also suits their own interests, but the acid test is what happens when complying with a (voluntarily adopted) code of ethics might cost you or your business money. 'Many companies today are writing codes that aren't worth the paper they're written on. . . . Ethics is about what you do not what you say' (Carmichael in Sasseen, 1993: 31). By itself, adopting a code of ethics is not a guarantee of social or environmental responsibility. This requires a genuine ethical concern and a sense of moral duty. This brings us back to Kant and his claim that having once recognized and voluntarily adopted an ethical principle (a categorical imperative), we can't simply ignore it or make exceptions to suit ourselves.

As we have argued, this does not necessarily make codes of ethics superfluous. There are undoubtedly many working within the tourist industry who have developed or adopted such codes because of genuine ethical concerns, concerns which they try to embody in their business practices. Their influence, along with the threat of legislation and social and consumer pressures, means that other, less scrupulous businesses may also decide to comply with such codes. At their best, they can provide clear normative frameworks that allow us to evaluate tourism operations and developments, frameworks that often take cognizance of others' rights and our duties. At worse, they are smoke and mirrors, superficial and abstract statements of apparent good intent but actual bad faith on the part of at least some signatories. As Kant argued, because it applies to everyone, morality is a public affair; one has to be able to state one's ethical principles openly. 'Morality means being fit to be seen. . . . Publicness is already a criterion of rightness in moral philosophy' (Arendt, 1989: 49). The problem with relying on codes of ethics arises from the tendency of many businesses to interpret publicness (moral openness) solely in terms of (self-serving) public relations and publicity. When this happens, then justice is far from being served, and it is to issues of justice that we now turn.

5 From social justice to an ethics of care

Tourism's strongest critics present a bleak analysis of a modern Western industry that, following the worst excesses of an imperialist past, continues to generate inequitable relationships. From such perspectives, Kaplan (1996: 63) argues, tourism 'arises out of the economic disasters of other countries that make them "affordable" or subject to "development", trading upon long-established traditions of cultural and economic hegemony, and, in turn, participating in new versions of hegemonic relations'. As Neale (1999: 227) points out, many within the tourism industry would reject such claims outright. But however beneficial tourism developments can prove, there can be no doubt about the need for some kind of protection for those peoples and environments most at risk from inappropriate forms of development. As Chapter 4 argued, despite their philosophical and political difficulties, rights and codes of practice seem to offer some such protection when they are properly enforced. For example, Simmons documents recent attempts to guarantee the intellectual property rights of those Aboriginal artists and communities whose work has been appropriated to promote the Australian tourist industry without attribution or financial recompense. 'In Australia', Simmons remarks, 'the tide has turned recently in favour of moral rights legislation', and new Copyright Amendment and Aboriginal Heritage bills will aim to 'protect both the cultural integrity of the artist's work as well as clan designs, symbols, and group ownership' (1999: 428–429). Rights, though, are only one part of a more general dialogue about the importance of social justice, and it is to this wider debate we now turn. We then briefly examine ethical discourses emphasizing 'care' and 'difference' that are more or less critical of the abstract notions of justice that tend to dominate modern ethical theory.

Social justice

Justice is the first virtue of social institutions, as truth is of systems of thought. . . . Each person possesses an inviolability founded on justice that even the welfare of society as a whole cannot override. . . . Therefore in a just society the liberties of equal citizenship are taken as settled; the rights secured by justice are not subject to political bargaining or to the calculus of social interests . . . truth and justice are uncompromising.

(Rawls, 1973: 3–4)

This quotation, taken from John Rawls's influential *Theory of Justice*, brings together a number of the ethical frameworks we have previously addressed. It is clearly critical of utilitarian approaches (Chapter 3), calls specifically upon the idea of inviolable human rights (Chapter 4), and reintroduces aspects of the virtue theory of Aristotle (Chapter 2). We saw that Aristotle thought of virtues as personal characteristics or dispositions that we should aspire to in order to further our well-being and happiness in a community of fellow humans. These virtues were all described in terms of a 'golden mean', a balance between extremes of behaviour that would otherwise prove socially disruptive, as *courage* was the virtuous mean between *cowardice* and *rashness*. For Aristotle, justice was the complete or sovereign virtue because it was the key to maintaining a harmonious social balance across the entire community; it was an over-arching public (rather than merely private) virtue. 'In justice is summed up the whole of virtue' (Aristotle, 1986: 173).

Like Aristotle, Rawls recognizes that a lack of justice threatens the well-being and happiness of the entire community. He also borrows from Aristotle the idea of 'justice as fairness' (Rawls, 1973, 2001). When people complain of an injustice they usually mean that they have not been treated fairly, equitably, or have been exploited to benefit an already privileged group. For example, Hall (1997) argues that the key ethical issue with sex tourism (tourism where the main purpose or motivation is to consummate sexual relations (Graburn, 1983)) in South-East Asia is not issues surrounding the morality of prostitution itself but the gender and economic inequalities that initiate and maintain this exploitative relationship:

> The sexual relationship between prostitute and client is a mirror image of the dependency of South-east Asian nations on the developed world. The institutionalized exploitation of women within patriarchal societies . . . has been extended and systematized by the unequal power relationship that exists between . . . host and advanced capitalist societies.
>
> (Hall, 1997: 119; see also Enloe, 1990;
> Stonich *et al.*, 1995)

This situation is reflected in other parts of the world where behaviour that the powerful classify as a holiday 'romance' seems suspiciously like a form of sex tourism from another perspective (Herold *et al.*, 2001). In such situations justice demands a change in this unequal power relation and that this change be reflected in current social institutions.

Justice, then, seems to be about the fair distribution of power, goods, and so on within and between societies. There are of course other, non-distributive kinds of justice. Aristotle also made a widely accepted distinction between distributive justice and retributive justice – that is, giving someone a just or fitting punishment (retribution) for their crime. As far as tourism development is concerned, we, like Rawls, can focus our attention almost entirely on distributive justice and its relation to social institutions. (Though many people who have felt exploited, like the Aboriginal communities and artists mentioned above, might also be seeking some form of retribution for past wrongs.) The question that we need initially to address is, then, 'What counts as a fair distribution?'

This is not as easy as it might sound. As children, pretty much all of us have argued at one time or another that we didn't get a fair share of the cake, that someone else had a larger or more attractive slice than ourselves. But such minor squabbles pale into absolute insignificance when we look at the worldwide distribution of basic necessities like access to food, medicine and education. As Gorringe (1999: 62) notes, the 1997 United Nations Human Development Report estimates that 'the share of the poorest 20 per cent of the world's people in global income now stands at a miserable 1.1 per cent, down from 1.4 per cent in 1991 and 2.3 per cent in 1960. It continues to shrink.' Consequently, infant mortality in a region like sub-Saharan Africa is much higher than average and people living in this region have a life expectancy of only 52 years. This compares to the 77 years that we can expect as residents of nations where per capita income is above $12,210 (World Bank, 1994). In other words, such inequality extends to life itself.

An obvious response is to argue straightforwardly for equity: for equal shares and benefits. We could represent this in terms of a chart as in Figure 5.1. But this is not necessarily going to be regarded as fair by all concerned. Some people might argue that they deserve a larger share because they have made a proportionally greater contribution to the total amount of benefits available, perhaps in terms of wealth creation or hard work. Others might make a case for extra resources on the basis that their special circumstances mean that their needs are greater, as might be the case with those with serious illnesses requiring extensive medical treatment. Both have a point, and it seems that a naïve or rigid form of absolute egalitarianism may be far from just, since we ought to take at least some of these claims into account.

This does not mean that the idea of equality is necessarily wrong in principle; it simply means that we need to distinguish between justifiable and unjustifiable claims to extra benefits. The World Health Organization states:

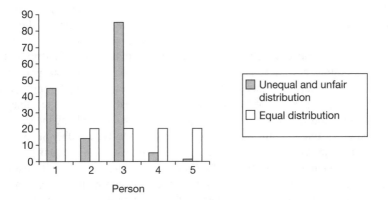

Figure 5.1 Two possible distributions of social goods

Differences in the distribution of burdens and benefits are justifiable only if they are based on morally relevant distinctions between persons . . . the equitable distribution of the burdens and benefits . . . raises no serious problems when the intended subjects do not include vulnerable individuals or communities.

(WHO in Schrader-Frechette, 1999: 77)

But this does raise the spectre of having to decide what counts as a morally relevant distinction. In limiting its concerns to vulnerable individuals and communities the WHO seems to run counter to free-market solutions which tend to emphasize the need to reward (and the 'moral' qualities of) the entrepreneur who creates innovative market opportunities and takes financial risks. Capitalists are willing to see quite extensive differences in wealth and power between individuals and communities on the grounds that 'enterprise' deserves to be rewarded.

Opponents of the excessive inequalities generated by capitalism have, on the other hand, tried to diminish or limit the gap between rich and poor, powerful and powerless. They are generally less happy to regard 'business enterprise' as deserving the kind of moral recognition that is used to justify excessive financial rewards. Indeed, many follow Karl Marx in arguing that capitalists are the moral equivalent of parasites, since their profits arise from exploiting the wage-labour of others. However, Marx does not, as is popularly believed, regard a distributive egalitarianism as either necessary or sufficient to characterize a fair society. The real needs of people – who in so far as they are individuals are always different (that is, unequal) – cannot be met by systems which treat people as standardized abstract elements and parcel up benefits accordingly. The famous communist slogan 'from each according to his abilities, to each according to his needs' (Marx, 1974: 347) is a call not for formal equality but for community, which is something entirely different. For Marx, a simple distributive equality does not go far enough. In a just society those capable of contributing more should do so to help those in less fortunate positions than themselves; the just society is about matching needs and abilities so that everyone benefits. In Zimbabwe the Parks Department has tried to resolve a similar issue by introducing a three-tier entry charge for national parks. Local people pay the lowest rate, regional tourists (those from South Africa) pay a higher entrance fee, and international tourists pay the highest fee. This is intended to take account of the fact that international tourists are much wealthier and more able to pay a higher entrance fee than poorer local people, who in any case pay for the upkeep for national parks through national taxation. The managers of national parks in Zimbabwe have noted that some international tourists complain, claiming that everyone should pay the same costs because everyone gets the same experience (see Duffy, 2000b). Supporters of the layered entrance fee system argue strongly that it takes account of the very different economic circumstance of local people. It allows them to have a proper right of access to their own material heritage: if everyone had to pay the same fee as the international tourists, very few local people would be able to visit the national parks.

Issues of justice and equality are clearly raised by any discussion of tourism development in the South. For example, Pattullo (1996) points out that very often, tourists misinterpret what they see as the laid-back attitude of tourism industry workers in the Caribbean. At best, tourists regard it as a quaint cultural form and an indicator of the relaxed outlook of local people; at worst, they judge it to be slothful and uninterested behaviour that can explain and justify why that culture remains poor and under-developed. Pattullo argues that in fact it echoes the passive resistances practised by slaves in the Caribbean context. As such, it interlinks with current issues of inequality between North and South that mirror the inequalities of the slave–master relationship. A more obvious example, but one that presents a complex ethical problem, is the issue of illegal tourism workers. In many tourism industries around the developing world, illegal migrant workers form the lowest-paid sector and they experience the worst working and living conditions. This is certainly the case in Belize, where illegal Guatemalan migrants work as chambermaids in a bid to find a better life in a neighbouring state (Duffy, 2002: 47–70). However, the ethical problem is encapsulated by debates about whether the illegal migrant worker is better off working in such poor conditions, which are after all better than those experienced in their home society since they have left behind economic poverty and political violence. Should such issues play any part in debates about what people deserve and how wealth is distributed? In Cohen's study of indigenous management of a tourism-related industry, even locally based projects resulted in benefits for some and not for others. Despite the economic achievements of some textile producers in Guatemala, overall, such developments had served to exacerbate socio-economic inequality (Cohen, 2001; see also Brown, 1992; Craik, 1995).

Whatever one's opinions on such issues, all justice theorists whether on the left or the right agree that some kinds of 'inequalities' are necessary and defensible in a fair society. Rawls's theory is meant to address what form these distributive inequalities might take, given the obvious lack of agreement about substantive questions about who deserves what and why. Rawls, like Aristotle (and for that matter Marx), regards justice as a political principle, as a necessary pre-requisite for genuine social solidarity. If individuals are to live and work together harmoniously, then they need to know that the society they will be part of is organized fairly. As Chapter 4 showed, early political theorists like Locke argued that societies could be thought of as originating in a kind of 'social contract' between otherwise isolated individuals, a contract that set out people's rights and responsibilities in return for their giving up their individual freedom. Rawls wants to ask what principles might underlie the apportioning of rights and respon-sibilities and the distribution of benefits and burdens in a 'contract' which at least in theory everyone would be happy to agree upon. He seeks to 'present a conception of justice which generalizes and carries to a higher level of abstraction the familiar theory of the social contract as found, say, in Locke, Rousseau and Kant' (Rawls, 1973: 11). If we could outline these contractual principles, we would also be outlining the principles of a just society.

Of course, Rawls does not think that such a contract actually exists; he merely wants us to engage in a kind of thought experiment whereby we try to discover what these principles might be. This experiment has to take place at a fairly abstract level because, as we have seen, when it comes down to the particulars of distributive justice people disagree very strongly. Rawls argues that such disagreements arise because people's views are biased by their actual position in contemporary society. In other words, a person who has made millions through instigating new developments in sex tourism is likely to have a perspective on the 'fair' distribution of revenues that is different from that of a person who has been forced to work in the same industry through poverty. You do not often hear millionaires complaining that it is not fair that they have so much money! The problem is, then, how to avoid such personal biases entering into the debate.

Rawls's solution is ingenious. He asks us to imagine that the individuals who are to discuss the constitution of a just society are completely removed from the realities of their everyday existence. Their discussion will take place behind a 'veil of ignorance' that shields the parties concerned from any knowledge about their previous existence and any inequalities in their social positions. They now occupy what he refers to as an 'original position' of equality. This is a 'purely hypothetical situation' where 'no-one knows his place in society, his class position or social status, nor does anyone know his fortune in the distribution of natural assets or abilities, his intelligence, strength and the like' (1973: 12). This, says Rawls, removes those morally arbitrary sources of bias that would otherwise make people argue for a society based on principles that might ensure that they continue to prosper at others' expense. If 'a man knew that he was wealthy, he might find it rational to advance the principle that various taxes for welfare measures be counted unjust; if he knew that he was poor, he would most likely propose the contrary principle' (ibid.: 18–19). Similarly, people's natural endowments, such as intelligence and strength, are largely chance inequalities, accidents of birth that no one can be said to 'deserve', and so, Rawls argues, they should not be allowed to distort distributive outcomes.

In this imaginary scenario, then, the millionaire knows nothing of his riches and the sex worker knows nothing of her role; they are on an equal footing. Being entirely ignorant of their position in society, they, like all rational people, would, Rawls argues, want everyone to have certain basic guaranteed liberties in place. Only in this way could they ensure that there would be a minimal safety net in case they found themselves in a disadvantaged position on re-entering society. Beyond this, though, they would accept inequalities in certain special circumstances, most notably when these inequalities might work to everyone's advantage. For example, some might argue that in a strictly egalitarian society there might be no incentive to work hard or to innovate since any of the benefits from this extra work would not accrue directly to the person concerned. If this is true, then allowing some differences in benefits might provide the kind of incentive that would increase the benefits available to all (the size of the cake). Since everyone benefits in such a situation, Rawls argues that we would all think this system fair. Thus Rawls posits two principles of justice arising from this thought experiment.

First: each person is to have an equal right to the most extensive basic liberty compatible with a similar liberty for others.

Second: social and economic inequalities are to be arranged so that they are both (a) reasonably expected to be to everyone's advantage, and (b) attached to positions and offices [equally] open to all.

(1973: 60)

Rawls further adds to the second principle the rider that inequalities should be 'to the greatest benefit of the least advantaged' (ibid.: 83). It is worth examining the rationale for these principles and their implications in a little more detail.

Rawls is trying to avoid imposing a particular model of a morally good life on people. He recognizes the diversity of ethical opinion but argues that this means that every person would want to be free to frame their own ideals, to decide on their own priorities. This in turn means that people would want the freedom to choose, and since we do not know what position in society we actually hold, we would want to ensure that everyone had these basic liberties/rights. For this reason, principle 1 (the concerns of liberty) always trumps principle 2 (concerns about efficiency and welfare). The second principle ensures that equality is the norm unless it can be shown that inequalities work in everyone's favour. It also ensures that there is equality of opportunity in terms of there being a level playing field for access to those positions of power and wealth that might exist. We could thus amend our graphical representation of absolute equality to fit with Rawls's arguments as shown in Figure 5.2.

However abstract Rawls's argument might initially seem, it provides us with a model of distributive justice that acts as a damning indictment of those examples of economic development which would remove people's basic liberties and mercilessly squeeze the poor for the benefit of the already rich. Yet, as many commentators have pointed out, it is not accidental that the picture of an ideal society that emerges from Rawls's deliberations is actually little more than a

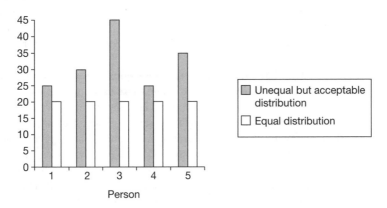

Figure 5.2 Rawls suggests that this kind of unequal distribution is acceptable because everyone benefits more than in a situation of absolute equality

liberal version of contemporary capitalism and is largely compatible with US legal and democratic institutions. For example, his

> ideal scheme . . . makes considerable use of market arrangements [because] it is only in this way . . . that the problem of distribution can be handled as a case of pure procedural justice. Further we also gain the advantages of efficiency and protect the important liberty of free choice of occupation.
>
> (Rawls, 1973: 274)

But this raises all kinds of questions about what kind of basic liberties/rights people actually want (see Chapter 4) and whether a government-regulated market economy is really the best way to achieve them.

It would certainly seem surprising to suppose that the abstract deliberations of people entirely removed from all cultural influences, as those in the original position are, would arrive at a political system so strangely familiar. But Rawls does not think that everyone who approaches the issue of social justice without personal bias would automatically come to the conclusion that a variation on the American Way is the only or best solution. Indeed, he has since argued that he was not trying to provide a universally valid moral framework, a 'comprehensive moral doctrine' (Rawls, 2001: xvii) for all cultures, but was aiming merely to lend theoretical support to a specific (liberal) 'political conception of justice for a constitutional democracy' (Rawls, 1992: 186). In other words, he admits that his arguments are only intended for, and will only appeal to, people who already share certain conceptions about the role of the modern state and its institutions. This is not necessarily as limiting as it might sound. To the extent that some of these same political conceptions have gained near-global legitimacy, his theory might still be regarded as providing a possible framework for thinking about social justice across many parts of the world. Indeed, we will discuss some of the issues surrounding distributive justice in terms of the ideal of ethical tourism associated with Zimbabwe's Communal Areas Management of Indigenous Resources (CAMPFIRE) programme in Chapter 7. Some of the limitations of Rawls's model become apparent when we look at cases of tourism development involving radically different cultures (see Butler and Hinch, 1996; Van den Berghe and Flores Ochoa, 2000).

But there are other challenges to the presuppositions behind Rawls's thought experiment that are not so easily dismissed. Rawls asks us to imagine the outcome of a rational debate between unbiased individuals about how to order society fairly. But in order to eradicate bias, Rawls has removed almost everything about these people that makes them recognizably individual. They are capable of thinking in abstract terms about their future self-interests, but everything that might have been part of their self-identity – their personal history, their physical characteristics, their social background, and so on – has been excised. This is why Michael Sandel says that Rawls's account of the original position is predicated on a notion of an 'unencumbered self'. The people we are asked to imagine are not Catholics, atheists or Buddhists, not clever or stupid, strong or weak, just bizarrely 'independent' selves whose 'identity is never tied to [their] aims and attachments' (1992: 23). But, Sandel argues, it is precisely these aims and attachments, these values, cultural differences and social relations, that make us recognizable individuals in the first place. The protagonists behind Rawls's 'veil of ignorance' agree only because there is no real difference between them; they are to all effects and purposes not just equal but identical.

Sandel is arguing not just that those occupying Rawls's original position seem unrealistically detached from real people, but that the premises of his thought experiment are wrong. Ethical values are an intrinsic part of our self-understanding and individuality. We cannot simply be separated from them since they are a constitutive part of who we are. What is more, such values are not just personal preferences for one state of affairs over another but are ultimately derived from our social and historically mediated inter-relations with others. To argue as Rawls does that we can understand people as 'self-originating sources of valid claims' (Rawls in Sandel, 1992: 20) is to risk falling back into a form of moral subjectivism (see Chapter 1) where ethical values are reduced to individual desires. Rawls's framework prioritizes the question 'What is right?' But this question makes no sense if we address it only to Rawls's abstract, characterless individuals because all real individuals already hold certain communally derived values, certain ideas of the good that influence their ideas of justice (see Box 5.1). So the question is, 'Can we really step back from the particular values we have and change them for new ones, or are we rather made the very people that we are by the values that we endorse, so that such detachment is impossible?' (Mulhall and Swift, 1993: 11). This question has been at the heart of subsequent debates between liberals like Rawls and so-called communitarians like Sandel, Alasdair MacIntyre and Charles Taylor, who emphasize the inter-relations between values, selves and society. Box 5.1 puts these rather abstract debates into a specific cultural and tourism setting.

Discourse/communicative ethics

As we have seen, John Rawls tried to provide an account of distributive justice that is widely, though (he later admitted) not universally, applicable. But if we do not have some such general account, then it seems we risk falling back into a form of moral relativism (see Chapter 2) in which we lack any 'objective' or unbiased way of comparing the competing notions of justice employed by different social groups (see Box 5.1). Unfortunately, as the 'argument around Rawls's notion of reflective equilibrium [the original position] illustrates, the burden of proof on any moral theorist who hopes to ground a conception of justice in anything more universal than the "settled convictions" of our political cultures is enormous' (McCarthy in Habermas, 1990: ix). Rawls was criticized both for the values he built into his account of the original position and for the values he left out. He built in and prioritized abstract notions of freedom, equity and rights that typify a liberal conception of justice and excluded all those specific concepts of *the good* that his communitarian critics insist are constitutive of real individuals and societies.

Box 5.1 Rawls's original position and real people: the case of the Sa

The Sa people of the Pacific island of South Pentecost, Vanuatu, retain many of their traditional values despite the influx of tourists who come to watch their spectacular *gol* ritual. (Indeed, they have actually employed tourist interests as a form of cultural and economic power to defend their traditional culture against other attempts to assimilate it.) Prefiguring contemporary bungee jumping, men and boys dive off stations on a high tower towards the ground beneath, hoping to be pulled up short by vines tied around their ankles. But this is no sport; it is an integral part of a traditional belief system linked to the harvest of Yams, which the Sa regard as a sacred food, and a reaffirmation of gender relations and ritual expression of male status.

> The symbolisms of the body of the tower and the social structure of the ritual, are supported by ideologies of gender expressed in *gol*. The male participants in the 'dive' are believed to display 'power' expressed as 'hot' force. (Yams, a sacred food, are also conceptually equated with male power as 'hot'.) In contrast, females – who dance beneath the tower – are equated with the 'cool' qualities of taro, a staple food.
>
> (de Burlo, 1996: 263)

Power in Sa society is not equally divided but linked to complex patterns of grades into which men gain entry by rituals based, among other things, on the sacrifice of pigs, payments for insignia, special dances and the acquisition of esoteric knowledge. 'Knowledge to construct and perform *gol* is also powerful' (de Burlo, 1996: 269). The *gol* ritual has become central to tourism development on the island and brings in significant amounts of tourism revenues. These revenues are distributed according to status, gender and one's role in the event itself (ibid.: 268), with women typically receiving $2–$5 and men $10–$20. In general, though, 'the ethos of the community demands a wide distribution of resources, similar to traditional exchanges. Virtually every person (except small children) receives a part of the tourist revenue' (ibid.: 268). A significant proportion is also set aside for group projects.

It is interesting, then, to think about this system and its associated values in Rawlsean terms (bearing in mind Rawls's later proviso that he means his theory to reflect an ideal of justice only for constitutional democracies rather than traditional social structures). Is the distribution of benefits just? Are people's liberties protected? Certainly it seems that there is a clear recognition that revenues should be widely distributed among the community's members. It might also be argued that those who take most risks to increase these revenues (diving from the highest stations) receive a proportionate reward for the risks they take. After all, without the spectacle presented by the 'land divers', tourist revenues would fall dramatically and the society's poorest members might suffer disproportionately. On the other hand, this is a highly stratified and gendered society, and roles and benefits are assigned according to tradition rather than being open to all, as Rawls's second principle demands.

Trying to make such comparisons also illustrates some of the problems with Rawls's original position, because he asks those abstract individuals behind the veil of ignorance to develop principles that are independent of any existing cultural ties. Yet the Sa idea of 'justice' is inextricably bound up with the specific features of Sa culture; it is not primarily about free and fair competition between individuals, or equity, but about fulfilling and reaffirming roles and expectations that are to a large extent pre-given by tradition. Thus, for example, to allow women (or for that matter tourists) an equal right to 'dive' would undermine the whole point of the tradition, which is precisely to reaffirm those social roles and hierarchies. This illustrates not only how Rawls's model, with its emphasis on equity, is not culture free but culture specific (as he now admits), but the communitarian point that ideas of justice are context dependent and inseparable from other culturally variable social norms like 'worth' and 'desert'. Such judgements are dependent upon precisely the kind of cultural information that Rawls deems biased and inadmissible (2001: 77).

Interestingly, the Sa themselves have little respect for the tourists who come to watch them, precisely because they regard them as the 'floating ones' (*aisalsalire*), lacking proper attachment to 'place and the sacred power it holds' (de Burlo, 1996: 273). Ironically, then, those abstract individuals occupying Rawls's original position would seem to epitomize everything the Sa regard as so 'pathetic' about tourists: they are extreme examples of free-floating selves to whom nothing is sacred. As such, they would hardly be ideal candidates for deciding what the Sa would count as just or fair.

Despite his original intentions, his account remains culturally relative (to modern Western societies) rather than universal and is premised on a false model of human life, societies and values.

A more recent attempt to rescue something of a universalist position centres around the work of the German social theorist Jürgen Habermas. Habermas too is critical of Rawls's reliance on abstractions like the original position and the 'unencumbered selves' he uses to found his theory of justice, but thinks Rawls was on the right track in considering justice as something that emerges from a discourse between interested parties. Indeed, Habermas goes further: he argues that there is something intrinsic to the act of communication itself, to engaging in a rational conversation with others, that assumes reciprocity between speakers. In a genuine conversation we have both to listen to and consider the ideas of others and to be willing to make our own ideas public so that others can criticize them from their own particular perspectives. We must also assume that the speaker means what they say and is willing, when asked, to provide further justification for their position. If we are not willing to do this, then we are not having a genuine conversation, but a knockabout more like a Punch and Judy show or a pantomime sketch where contrary opinions are simply bandied back and forth without resolution. The aim of a real conversation is to reach agreement, and 'the very act of participating in a discourse involves the supposition' that genuine consensus is possible' (McCarthy, 1984: 306).

Habermas argues that reciprocity, respect and the search for a resolution are necessary and universal features of all genuine human communication – that is, they hold for all cultures. This is not to say that all social communications actually follow this pattern; people do lie, distort each other's arguments, avoid answering difficult questions, and so on. But such 'conversations' are not *genuine* attempts to communicate with others precisely because of these faults. Rational discourse entails that we have respect for the position taken by others, giving them the room to explain their beliefs and values. We must also be willing to change our own perspective so that the outcome of any such discourse will be dependent upon the argumentative force of the various positions. The best argument will win out. This at least is what Habermas supposes would happen in what he refers to as an 'ideal speech situation', where extraneous factors, such as the power or prestige of certain speakers, that might otherwise carry undue influence, make people bow to pressure, or accept poorer arguments, are excluded.

Habermas's 'ideal speech situation' has several advantages over Rawls's 'original position'. The people involved are not unencumbered selves but real people with all their values intact and with full knowledge of their social roles and relations. They carry this social 'baggage' with them because, as communitarians argue, it actually forms part of their individual identities, and provides the basis for their different positions in any subsequent discourse. There is no artificial equality here, then, no attempt to put everyone in an identical situation removed from their everyday lives. It is, however, Habermas argues, a fact of life that whenever we engage in genuine conversation we have, by definition, to be respectful of others' arguments. In other words, 'the idea of impartiality is rooted in the structures of argumentation *themselves* and does not need to be *brought in* from outside' (Habermas, 1990: 76). This means that Habermas can, like Kant, claim that his ethics is genuinely universal in application; it must apply to all genuine arguments about values whatever language or culture they take place in. Indeed, he even presents his argument in terms of a reformulation of Kant's categorical imperative (see Chapter 4):

> From this viewpoint, the categorical imperative needs to be reformulated as follows: 'Rather than ascribing as valid to all others any maxim that I can will as a universal law, I must submit my maxim to all others for purposes of discursively testing its claim to universality.'

> (ibid.: 67)

The ideal speech situation, then, provides the opportunity for this discursive testing of our ethical claims.

But where might we find this ideal speech situation in real life? Certainly not in many aspects of the tourist trade, where, for example, tour guides often feel compelled to conform to client expectations and only engage in stereotypical conversations that further validate the visitor's idea or construction of a destination and its inhabitants (Hall, 1994: 174–182; McKercher, 1993: 12–14). The company boardroom with all its inherent inequalities and dependencies – 'I'd better say what they want to hear rather than what I really believe if I want that promotion' – also seems very far from an ideal speech situation. Perhaps, though, something like a public inquiry might at least be viewed as a 'communal' attempt to find the truth of a situation – to investigate, for example, the arguments for and against a new airport development. This contrasts with boardroom debates, which are often little more than a parade of competing, rather than co-operating, egos, each vying for favours in an atmosphere where decisions have often, in any case, already been taken behind the scenes. But the problem is, the more we think about it, the less likely it seems that we can ever create an ideal speech situation. Even public inquiries are often reduced to bun-fights where prestigious lawyers try to twist the words of their less experienced opponents. In this sense at least, the ideal speech situation seems a fiction as far removed from real life as Rawls's original position.

However, Habermas is fully aware of this. What he is really arguing is that whenever we enter into a rational debate, we *must* actually assume that something like the ideal speech situation holds. If it doesn't, then we can actually challenge the conclusions of the debate on the grounds that it was determined by some (irrational) external factor. For example, we might argue that the remit of a public inquiry was set too narrowly, that key witnesses were bribed, vital evidence suppressed, and so on. All of these involve a failing to live up to our rational expectations about what claims to be a genuine attempt to communicate ideas and achieve a consensus. The ideal speech situation is exactly what it says: an ideal embodying the requirements of genuine communication to which we can compare real situations so as to highlight their deficiencies.

Habermas's argument seems very appealing because it seems to make social justice both the outcome of and the basis for all genuine communication. The ideal speech situation both provides a procedural model for how to resolve ethical differences and embodies in itself the pragmatically necessary and universally present principles of communicative justice. In short (and oversimplifying issues somewhat), all we need to do when differences arise is to sit down and talk things through rationally. And, since Habermas claims that the normative features of genuine communication transcend (go beyond) particular cultural boundaries, this 'discourse ethics' should be especially useful to the kind of cases that emerge as a result of tourism development, globalization or, for that matter, in a multicultural society in general (for further discussion of tourism and globalization, see Duffy, 2002: 127–154; Wood, 2000).

Indeed, Dryzek (1999) argues that Habermas's position can be used to legiti-mize the conclusions that emerge from meetings of the citizens, non-governmental organizations, activists, and so on, who he claims form an emergent and *global* 'civil society'. The specific example he cites is the discussions that took place at the unofficial Global Forum which met alongside the official UNCED meetings at the Rio Earth Summit in 1992. If, as Habermas's (1995) work suggests,

> democratic legitimacy is to be found not in voting or representation of persons or interests, but rather in deliberation [then] . . . an outcome is legitimate to the extent that its production has involved authentic deliberation on the part of the people subject to it.
>
> (Dryzek, 1999: 276)

The relatively unconstrained deliberations at these unofficial meetings of minds might be seen as more democratic than the formal and limited discussions of the official event itself.

> Compared to the realm of states and their interactions, civil society is a realm of relatively unconstrained communication; its actions are not bound by reasons of state, by the conventions of diplomatic niceties, by the fear of upsetting allied or rival states, or (more important) . . . the financial markets.
>
> (ibid.: 278)

This, of course, has enormous implications for the way in which development issues can or should be decided. It suggests a model of participatory rather than representative democracy that is altogether more inclusive, that can cross national and cultural boundaries, and where the conclusions reached are much more open to argument. It might be described as 'anti-institutional' or even 'anarchistic', although Benhabib (1990: 18) argues that it is better understood as developing 'a normative and critical criterion by which to judge existing institutional arrangements, insofar as these current arrangements suppress a "generalizable interest"'. In giving a voice to interests and ideas that are otherwise suppressed, discourse ethics also seems to lend ethical and political support to developments such as community-based tourism projects. Such projects are designed to ensure that the local community is actively involved in decision making and management of tourism developments (Drumm, 1998; Lindberg *et al.*, 1996: 547–551; Moscardo *et al.*, 1996: 29–33; Place, 1991: 196–201; and see Chapter 7). As advocates of community-based tourism, such as Tourism Concern, argue, social justice is about more than just giving a fair share of profits back to the local community; it also requires the involvement and consent of the whole community.

There are, however, some theoretical and practical problems associated with a discourse ethics approach. If properly practised, discourse ethics would clearly be very time-consuming for those concerned. In this, and in taking decision making out of the hands of experts and committees, it clearly goes against the whole modernist and bureaucratic ethos that wants to set up rules, codes and regulations based on a need for ever-increasing efficiency. In this sense, communicative ethics may not sit well with business requirements or with contemporary governmental and international institutions, but then its advantage lies precisely in presenting us with an alternative way to proceed that puts social justice first. From a practical point of view the attempt to model our deliberations on an ideal speech situation can seem unnecessarily utopian. On the other hand, the fact that we cannot actually reach this perfect situation does not stop us employing it as a measure of the moral distance we still have to go. And, as Benhabib remarks,

> Far from being utopian in the sense of being irrelevant, in a world of complete interdependence among peoples and nations, in which the alternatives are between non-violent collaboration and nuclear annihilation, communicative ethics may supply our minds with just the right dose of fantasy.

> (1990: 20)

Another practical issue is that of who gets to take part in the discourse. Who gets to sit around the table to discuss the rights and wrongs of actual cases? If we think of tourism developments, then there will always be many parties involved, including the developers, the local populace, potential employers and employees, the tourists themselves, the business community in general, environmentalists, and so on. Can anyone and everyone pitch in, and, if so, should everyone really have an *equal* voice? One could argue that there are certain situations where it is actually right to give more weight to what some have to say over others on the basis of tradition, need or level of involvement. For example, we might want to consider more carefully the arguments of certain individuals like tribal or village elders, or those who have most to gain or lose from the situation under discussion. This, though, has its own dangers, because even where they seem to represent a view held by many, such power can actually be used to reinforce the hegemony of an elite group. For example, Van den Berghe and Flores Ochoa (2000) point out that while reverence for the Inca past (*Incanismo*) in Peru is widely held, providing an ideology of local pride and a marker of regional identity, it actually articulates and perpetuates a patrimonial society that disproportionately benefits wealthy families. In defence of discourse ethics, one might respond that we should construe these situations as deviations from an aspirational norm, as special cases requiring justification within the context of the ethical discourse itself.

Other objections centre on the theoretical presuppositions behind the ideal speech situation itself. One could argue that Habermas's claim that ethical discourse can and should aim to produce a consensus is open to challenge. First, as Benhabib argues, 'consent alone can never be the criterion for anything, neither for truth nor of moral validity' (1990: 12). The fact that a group have finally reached an agreed solution to their differences does not necessarily mean that the compromise is *ethically* superior to the speaker's original positions; it just means that they have effected a compromise. Indeed, it may actually mean that all involved have actually *compromised* their ethical values in the rather different sense that they could be held to have acted somewhat shamefully. As Chapter 1 argued, many ethical values are identifiable precisely because they are not the kind of thing one can 'trade', give up or compromise on without imperilling one's reputation for ethical *integrity*. This also links back to issues raised by communitarians, who might argue that ethical values are such an integral part of one's self and cultural identity that they cannot be abandoned or altered without changing the very nature of the individuals and communities concerned. Is discourse ethics, then, almost a form of rational coercion (McCarthy, 1993: 191) and its eventual endpoint a world devoid of, rather than respectful of, difference, a kind of bland consensus where everyone has the same values?

Some of Habermas's interpreters have tried to argue that this is not necessarily the case. Steven Vogel claims that Habermas makes a distinction between ethical values and moral norms and procedures. Habermas seems to rule out achieving consensus about actual *ethical* values, which are too varied and too strongly linked to our historical circumstances and self-identities, and argues that discourse ethics can actually resolve only 'moral' problems, which he defines as norms and rules that regulate interactions between people (see Vogel, 1996: 149). From a slightly different perspective Seyla Benhabib argues that we can adapt Habermas's perspective so that rather than seeing consensus as the endpoint of moral dialogue we regard discourse ethics as stressing the processes and importance of dialogue itself:

> We begin to ask not what all would or could agree to as a result of practical discourses to be morally permissible or impermissible, but what would be allowed or even necessary from the standpoint of continuing and sustaining the practice of moral conversation among us.
>
> (1990: 13)

Habermas himself seems to admit that there are limits to what can and cannot be resolved through discourse ethics, especially where radically different values are involved. The 'sphere of questions that can be answered rationally from the moral point of view shrinks in the course of the development toward multiculturalism within one society and toward a world society at the international level' (1993: 91). But if this is so, and differences in ethical values are not subject to resolution through discourse, then much of the impetus for developing communicative ethics as a universalist framework seems lost. This, together with arguments about the presumptions inherent in Habermas's idea of rational moral argument, forms the basis for both feminist and post-modern critiques, and it is to these we now turn.

Feminism and an ethics of care

It is crucial to Habermas's conception of discourse ethics that ethics is a debate between *rational* individuals. But that leaves open the question of what counts as moral reasoning. Habermas believes that this is a capacity that individuals acquire as they mature, and bases this claim partly on the work of psychologist Lawrence Kohlberg (Habermas, 1990). Kohlberg argues that children start life without any ethical concerns, interested only in their own self-fulfilment. Gradually they come to adopt the moral norms of the society around them, at first without reflecting on or criticizing them. Only later do they reach moral maturity, a stage marked by their being able to employ abstract moral principles, like those of equality, justice, and so on, across divergent cases – this is moral reasoning. From his experiments with boys and girls, Kohlberg claimed that girls were less likely to reach these latter stages, suggesting, at least implicitly, that they were less capable of moral reasoning.

Kohlberg's co-worker Carol Gilligan (1983) took issue with this interpretation. While agreeing that boys tended to speak with the abstract voice of 'justice', which Kohlberg labels a sign of moral maturity, she noticed that girls made complex moral judgements based on a detailed understanding of specific circumstances, contexts and personal relations. She argued that this 'contextual' voice is no less developed than that of abstract justice; rather, women speak in a *different* [moral] voice. Gender-specific experiences of childhood mean that girls are expected to emphasize the caring and relational aspects of ethics (because of their conventional roles in maintaining family and social relationships), while boys come to adopt the dominant masculine model of maturity as autonomous, rational and independent (Chodorow, 1978). In other words, children adopt forms of ethics that encapsulate the gender stereotypes of our patriarchal (male-dominated) society. Gilligan's feminist critique 'calls into question the values placed on *detachment* and *separation* in [Kohlberg's] developmental theories' (Gilligan, 1994: 213) and in modern Western society in general.

> Where Kohlberg equates ethical maturity with attaining the ability to remain 'objective', *to stand back* and apply abstract rules and principles to moral problems, Gilligan emphasizes the ethical subject's *closeness* and *attachment* to others. Her moral agent is not an egocentric automaton but a relational self, that is, a person whose relations to others are *constitutive of* rather than *incidental to* her identity. Rather than developing a moral theory with universalistic pretensions, Gilligan advocates paying attention to the differing situations and contexts in which we actually concern ourselves with others.
>
> (Smith, 2001a: 174)

There is an interesting parallel here between Gilligan's critique of Kohlberg's account of the process of human development and critiques of the dominant 'modernization' theory of social and economic development (see Chapters 3 and 4). Gilligan points out that 'one definition of the word [development] presupposes a linear model of progression to a better form [but] [a]nother definition proposes an unfolding, the realization of a fuller potential' (Heckman, 1995: 14). Heckman describes how 'Gilligan's challenge to the linear concept of development and its masculinist bias is a function of her more fundamental challenge to the paradigm of the autonomous self that grounds that theory' (ibid.: 15). The 'autonomous self' which, as earlier chapters have shown, forms the basis for almost all modernist ethical theories is, Gilligan claims, not gender (or culture) neutral. Rather, it embodies a particularly masculine and modernist ideal (Lloyd, 1984) that inevitably leads to the under-evaluation of those aspects of ethical *relationships* conventionally associated with women. In a similar vein, dependency theorists have argued that socio-economic '"development" tends to be short for the Western development model. The perspective remains linear, teleological, ethnocentric' (Pieterse, 2001: 24). They too have rejected the idea of a single progressive line of developmental stages towards an endpoint that simply recapitulates the form of modern Western society. They too want to recognize the importance of different voices and different, and potentially more fulfilling, patterns of development that run counter to the dominant model (Crocker, 1991).[1] This masculinist bias is evident in much of the tourism literature as well. Johnston suggests that tourism research reproduces hegemonic, disembodied and masculinist knowledges in line with broader social science research that has been reliant on the mind–body dualism that privileges the mind over the body. The explicit recognition and inclusion of gendered, sexed and sexualized bodies in tourism research provides a narrative that subverts the masculinism of tourism discourses (Johnston, 2001).

Gilligan's argument that 'the moral problem arises from conflicting responsibilities rather than competing rights' means that she regards ethical reasoning as 'a mode of thinking that is contextual and narrative rather than formal and abstract' (1983: 19). Ethical thought is a kind of 'storytelling' that has to be grounded in the complex practicalities of everyday life; the narrative must tease out the differing forms of ethical relations between real people. It cannot make the kind of grand generalizations beloved of traditional ethical theory (see Box 5.2). This is not to say that an 'ethics of care' is necessarily superior to 'justice' ethics (at least when broadly conceived), as some of Gilligan's successors have on occasion seemed to suggest (Noddings, 1984; Ruddick, 1989). In her more recent writings Gilligan argues that the impartiality associated with conceptions of justice and the partiality associated with an ethics of care are both important, providing a necessary counterpoint to each other's excesses. Unfortunately, it is far from clear quite how these distinct approaches might be combined to form a harmonious whole, a new ethics of justice *and* care (but see Heckman, 1995; Held, 1995).

Box 5.2 Tourism and an ethics of care: the ethics of tourism research

Researching tourism often requires the building of a personal relationship between the interviewee and interviewer. This has to be seen against a backdrop of a complicated power relationship between myself (R.D.) as a researcher and the people whom I interview. I have to be careful and sensitive about the fact that I am a white Western researcher with access to finance to pay for research, that I work in a university based in a former colonial power, that I am a woman, and so on. All of these factors (and more) have a critical bearing on the ways that I conduct tourism research and relate to the interviewee. There are many ways in which my own research might have impact on the lives of those to whom I speak. In Zimbabwe I was once shown some documents that detailed who was involved in the ivory trade and in poaching rings. Yet revealing this through my research would have compromised the personal safety of the source (and in fact would have ensured that they were in danger of being killed).

Academic disciplines produce ethical guidelines about interviewing, but even these supposedly hard and fast rules have to be interpreted contextually. As Gilligan might argue, in order to take decisions about when it is right to use information that has been passed to me confidentially and anonymously, it is important to understand the particular context of the research. Rather than employing a kind of Kantian absolutist rule which might argue that I should *always* reveal evidence of lawbreaking to the appropriate authorities, or *always* adhere to my promise of anonymity to the interviewee, I had to think through many other issues in considerable detail. The information might have helped expose some of those responsible for poaching and perhaps have impacted on the ivory trade, yet I also clearly had to take into account issues about the interviewee's safety, about presenting a truthful account in my academic work, and so on. Gilligan suggests that a caring relationship is one in which we develop understanding and concern for the others involved. But there are many people (and animals) involved here, and even the interview relationship allows only a brief period in which to develop any kind of understanding of their position and of the full implications for them of what they are telling me. This emphasizes the importance and difficulty of trying to be sensitive to particular contexts, rather than regarding my role as an interviewer as someone there to take an entirely 'objective' and distanced stance. In every case it is important to treat the interviewee as a unique individual who is always much more than a mere *source* of information. In this particular case I decided not to use the information. Instead I wrote about ivory trading in very general terms and in a way that did not place interviewees in jeopardy. After all, I can leave Zimbabwe at any time, while the interviewees have to continue to live and work within that society.

My solution to such research issues does not provide the perfect pathway through this complex ethical problem, but it does perhaps instead indicate the ways that individual researchers have to negotiate difficult circumstances and make personal (and moral) judgements about their research and publication strategies.

Difference ethics

There are, as David M. Smith (2000: 83) points out, 'obvious links between the ethic of care and communitarianism': both emphasize relational rather than autonomous notions of identity, both imply partiality, and both 'appear to rely heavily on the proximity of the persons concerned, to make their morality work' (84). In all these senses Gilligan's feminist approach is, like communitarianism, critical of the dominant modernist traditions in ethics which try to ground their account of morality on the basis of some universally shared feature or features, like self-interest, rationality, and so on. Modernist ethics tries to counter the clear disparities within and between cultures by reaffirming our common humanity and appealing to our enlightened self-interest, as in ideas of a social contract or in Rawls's original position. It argues that despite the obvious diversity of subject positions people occupy (in terms of culture, wealth, power, and so on), if we stand back and think about things in more abstract terms, then we will recognize that so far as all morally relevant criteria are concerned, we are identical. In other words, the key to the modernist moral enterprise is to base arguments on some notion of a shared *identity*, rather than, for example, virtue ethics, where virtues are associated with particular subject positions (see Chapter 2). This is why Smith regards both communitarianism and an ethics of care as reverting to pre-modern forms of 'parochialism' (ibid.: 86). This is, however, a mistake, because an ethics of care actually has much more in common with those approaches often labelled 'post-modern', such as the 'difference ethics' developed by philosophers like Emmanuel Levinas (1991; Hand, 1989) and Luce Irigaray (1993).

Post-modernism has been defined in many ways. Some speak of post-modernity in terms of a new era, as entailing a cultural shift as radical as the movement from feudal to modern societies. Some simply regard it as an extension and more extreme expression of modernism itself. Frederick Jameson (1991) refers to it as the 'cultural logic of *late* capitalism'. Jean-François Lyotard, author of *The Postmodern Condition* (1984), defines it as a cultural condition in which we are forced to bring modernity's prevailing assumptions into question.

> 'Generally perceived as positivistic, technocratic, and rationalistic, universal modernism has been identified with the belief in linear progress, absolute truths, the rational planning of ideal social orders, and the standardization of knowledge and production.' Postmodernism, by way of contrast, privileges 'heterogeneity and difference as liberatory forces in the redefinition of cultural discourse.' Fragmentation, indeterminacy, and intense distrust of all universal or 'totalizing' discourses (to use the favoured phrase) are the hallmark of postmodernist thought.
>
> (Harvey, 1990: 9)

The anti-foundationalism typical of post-modernism is often regarded as expressing (and in some cases supposedly celebrating) rootlessness in a world dominated by the constantly accelerating circulation of disconnected values, symbols, narratives, etc. Thus Jean Baudrillard (1993: 54) proffers an account of a hyper-modern world where '[t]ravel is a necessity and speed is a pleasure . . . where movement alone is the basis of a sort of happiness'. Here nature and culture alike become subsumed in a system of manipulable signs, an infinitely flexible currency of exchange without any grounds whatsoever (Smith, 2001a). In this sense, and as Baudrillard's own account of his American tours illustrates, tourism is itself 'prefiguratively postmodern' (Urry, 1997: 87). It is predicated on the groundlessness of both the traveller and of the constantly manufactured significance prescribed to the cultural and natural heritage they consume. The nature of the post-modern tourism experience is concerned with gazing on a spectacle with uncertain origins. The identities of destinations are invented and reinvented to create powerful images and representations of peoples and places. In this spectacle the real and the hyper-real can collapse into one (Ateljevic and Doorne, 2002; Ryan *et al.*, 2000). It is as though, in a post-modern landscape, we can no longer distinguish between Disney World and the real world. Indeed, to all effects Disney World has become the real world.

> Accounts of leisure in postmodernity stress the decomposition of hierarchical distinctions between high and low culture; irresistible eclecticism and the mixing of codes; the pre-eminence of pastiche, gesture, and playfulness in social interaction . . . the depthlessness and transparency of activities. . . . Postmodern leisure is, as it were, existence without commitment.
>
> (Rojek, 1995b: 7; see also Cohen, 1988; Waitt, 2000)

We will return briefly to some of these issues in the next chapter. But if post-modernism does mean 'existence without commitment', this might seem to imply the end of ethics altogether. That is not necessarily so, but it does entail 'the rejection of the typically modern ways of going about its moral problems (that is, responding to moral challenges with coercive normative regulation in political practice, and the philosophical search for absolutes, universals and foundations in theory' (Bauman, 1993: 4)).

While Levinas and Irigaray do not refer to themselves as 'post-modern' thinkers, they both reject totalizing discourses, such as those of 'rights' or 'utility', and emphasize the ethical importance of *difference* rather than *identity*. Ethics is, on Levinas's account, a 'face-to-face' relationship with others through which we come to recognize and prioritize their needs rather than treat them instrumentally as beings through which we can further our own ends or desires. 'I *can* see another as someone I need in order to realize certain wants of mine. She or he is then a useful or enjoyable part of my world, with a specific role and function' (Peperzak, 1992: 19) – say, a hotel porter. However, this is not an ethical relation; an ethical relation requires that I pay heed to the specific presence of another. This person 'come[s] to the fore *as other* if and only if his or her appearance breaks, pierces, destroys the horizon of my egocentric monism' – that is, if I stop seeing everything around me as part of *my* world or everyone else as someone there to satisfy *my* desires.

For Levinas and Irigaray, an *ethical* relation to others requires us to respect their *difference* from ourselves, to refuse to make them subject to *our* desires, compliant with *our* expectations, or forced to speak with *our* words. Ethics resists the desire to assimilate difference, to possess the Other, to make them fit within the limits we might want to prescribe. The ethical is a kind of relationship in which the 'Other' is made manifest in its irreducible difference to ourselves. In Levinas's terms, ethics is a relation of *infinity* rather than *totality*. It does not seek to monopolize, sublimate or account for the Other within our own circumscribed horizons but recognizes that that which is other to ourselves will always, of necessity, *transcend* (be in excess to) these limits (Smith, 2001b).

This ethical approach has obvious relevance for tourism. Critics of colonialism have pointed out that tourism has often constructed a very *unethical* conception of the 'Other', as something or someone alien, mysterious and exotic (Said, 1978). This very different process of 'Othering' utilizes a series of binary oppositions or dualisms where each term is used to define its opposite number: Western and Oriental, male and female, white and black, rational and irrational/emotional, civilized and uncivilized, modern and primitive, and so on (Johnston, 2001; Plumwood, 1993; Taylor, 2001). Here, then, the subaltern term is simply something which helps to define its opposite as superior; the 'hosts' and their culture are simply a resource to help define the identity of the 'guests'.

> Places and peoples are constructed as signifiers of the Other to be consumed by tourists. . . . Tourist destinations as *sites* for tourists, and the people within them as *sights* for tourists, are frequently rendered Other by a tourist industry that has developed an unsigned colonial and gendered hegemony in the form of a set of descriptors for constructing and representing 'Tropical Paradise'. . . . The people within these landscapes are frequently portrayed as passive but grateful recipients of white explorers from urbanized and industrialized countries searching for their authentic origins.
>
> (Aitchison, 2001: 135, 137)

A 'difference ethics' requires us to have a very different kind of relation to the Other. In Luce Irigaray's terms, ethics is a passionate relation in which *desire* for the other is tempered with *wonder*. Where desire alone would possess the Other, define her or occupy her, reducing her to an economy of the Same, '[t]he "object" of wonder . . . remain[s] impossible to delimit' (Irigaray, 1993: 81).

> [W]onder goes beyond that which is or is not suitable for us. The other never suits us simply. We would in some way have reduced the other to ourselves if he or she should suit us completely. An excess resists . . . the other's assimilation or reduction to sameness.
>
> (ibid.: 74)

In other words, both Levinas and Irigaray characterize ethics as a relation to others that is intimate and yet seeks to *sustain* a respectful distance. We both desire and are touched by the Other's presence, yet our wonder at their being other than ourselves ensures that our touch remains gentle and our desire is not to consume or confine them but to conserve the uniqueness of their being (Smith, 2001b: 362). This, of course, is the very opposite of the dominant trend towards the increasing commodification of social relations discussed in earlier chapters.

Since difference ethics requires us to recognize and respond on a very personal level, this ethical relation to the other cannot be spelled out in formulaic terms, in codes or regulations or in the apportioning of rights and duties. Like an ethics of care, it requires us to be aware of the needs of others and to embody this awareness in our actions. It is, then, an ethic for the individual tourist rather than a means of regulating the tourist industry or tourism development. It suggests that tourists must strive to become aware of those around them as much more than a part of their leisure experience; they should respond to others on their own terms. The difficulty is that if post-modern theorists like Baudrillard are correct, our relationships, like our knowledge and values, are now all so ephemeral and superficial that it makes little sense to speak of an authentic ethical relation. It is to the question of authenticity and its ethical implications that we now turn.

6 Authenticity and the ethics of tourism

We concluded Chapter 5 by pointing to the central importance in recent ethical theories of a respect for difference, of a need to remain genuinely open to and help sustain what, for want of a better word, we might term 'otherness'. From this perspective the ethical relation is one that refuses to reduce another person, place or culture to that which we want it to be or that which might be useful for us. Nor should differences be regarded as something alien or exotic, as things that serve only as a complete contrast to what we regard as normal and normative. The 'Other' is not someone or something that should be reduced to a stereotypical or essentialized image that can be consumed to feed our already established notions of our self-identity. We suggested that this willingness to understand another person, place or culture for what they really are, not what we would like them to be, is a particularly apposite ethical model for individual tourists' relations to the cultures they visit. The problem is, though, that the very process of tourism development inevitably changes the self-identity, environment and society of those affected. As it does so it becomes increasingly difficult to know what others are *really* like, or to distinguish what is real and original from a commodified presentation designed to fulfil touristic expectations. If authentic experience might be thought of as the Holy Grail of ethical tourism, it often seems to be just as elusive.

By 'authentic' we usually mean that something is genuine and original, that it can be certified by evidence, or remains true to a tradition. The tourism industry relies heavily on the idea of authenticity, and tourists themselves often seek out what they define as an authentic local experience. The idea of 'real travel' is dependent upon a notion of a genuine local experience, which raises the issue of what is defined as traditional, original and local. This question of authenticity has become crucial to the literature on tourism development (Cohen, 1988; MacCannell, 1999; Pearce and Moscardo, 1986; Redfoot, 1984), partly because tourism and the heritage industry often seem to dissolve the boundaries between what is verifiably authentic and inauthentic or counterfeit. The tourism industry tends to provide its own definitions of the traditional or typical, which are negotiated through what is locally perceived as authentic and what tourists and developers view as key travel experiences.

What tourists usually see is the performative aspect of local cultures presented to visitors in a stylized, ritualized and palatable version. While this 'performed authenticity' is created, staged and carried out for external consumption, it is important to place it in the context of how and why the tourism industry defines and presents its version of the genuinely local. This marketing of a particular image of a nation-state, a region, a culture, an environment or landscape is imbued with ideas that are primarily, but not exclusively, intended to appeal to external tastes and Northern concepts of the exotic. The image of a destination is important to attract customers (tourists), and this creation by the tourism industry encourages the inhabitants of those destinations to recognize a caricature of themselves and then perform it for visitors (see Lanfant, 1995). This is especially significant in terms of North–South relations, because the arrival of Northern tourists on Southern shores raises particular ethical issues surrounding race, colonialism, economic and political inequality, and the commodification of local cultures. It does, however, also raise important questions about the appropriation of heritage in the North (Dicks, 2000).

There is also a philosophical aspect to this question of authenticity that is inseparable from the form taken by late modern, or post-modern, society and the manner in which tourism epitomizes and exports this increasingly disorientating and fragmented culture (see Chapter 5). Some theorists of late modernity have argued that the kind of authentic relationship to oneself and one's fellow citizens that seems to be a pre-requisite for ethics is hardly possible in a culture where nothing seems certain or fixed and our very identities are constantly being remade. We live a 'life in fragments' (Bauman, 1995) where our self-identities are increasingly being pulled apart and remade by the pressures of (post-)modern existence. Those inhabiting this (post-)modern world 'find themselves confronted with captivating, seductive and expansive options that allow people readily to exchange one identity for another' (Zimmerman, 2000: 123). (As we have discussed in previous chapters, tourists are not a homogeneous group, performing a single identity. Rather, there are multiple types of tourists (mass, eco, adventure, cultural, and so on) who perform different roles at various times during their vacation (see Cohen, 1979; Gibson and Yiannakis, 2002; McMinn and Cater, 1998; Wickens, 2002).) New technologies and the culture of consumption that allows us to buy and present new identities (explorer, hiker, adventurer) lead to a

> fragmentation of self-conceptions [which] corresponds to a multiplicity of incoherent and disconnected relationships. These relationships pull us in myriad directions, inviting us to play such a variety of roles that the very concept of an 'authentic self' with knowable characteristics recedes from view.
>
> (Gergan, 1991: 7)

This 'fragmented society' as Taylor (1991: 117) argues, 'is one whose members find it harder and harder to identify with their political society as a community'. People thus come to 'see society purely instrumentally' (ibid.) rather than ethically.

Perhaps, then, the tourist's search for authenticity in other cultures is a search for something lacking in their own, perhaps even a search for an illusive self-identity. Travel is, after all, conceptualized as escape, a means to explore other cultures, as broadening the mind, as a means of self-discovery and freedom, and as a period of hedonism or happiness (Dann, 1996a: 101–134; Krippendorf, 1987: 22–28). People travel to seek a temporary refuge from the cycle of everyday life, which revolves around work, home and free time (Krippendorf, 1987: 3–19). Tourists certainly have a number of different motivations for their choice of destination and type of holiday, including a desire to have novel experiences and to get away from it all (Gnoth, 1997: 286–300). But all these motives might be regarded as a felt need for respite from the exigencies of modern life, and/or as 'authentic' projects of self-discovery.

The ways that tourists conceptualize, define and describe their tourist experiences reveal that travel stories assist individuals in narrating self-identity. Those identities are then affirmed and contested through discussion of 'travellers' tales' (see Desforges, 2000; Elsrud, 2001; Galani-Moutafi, 2000; Murphy, 2001). Munt has suggested that even those most conscious of their ethical responsibilities, like the independent, environmentally oriented travellers who might be labelled ecotourists (see Chapter 7), are fundamentally self-interested. Munt's position is in many ways analogous to the arguments for moral scepticism presented in Chapter 1, arguing that such tourists are, in reality, ego-tourists who search for a style of travel that reflects their own perception of themselves as having an alternative lifestyle. Travel is all about enhancing and maintaining the individual's cultural capital (Munt, 1994a: 105–108). Whether or not this is entirely fair, it is certainly ironic that tourists only seem to bring their culturally derived uncertainties with them on their travels and their search for authenticity leads to the inevitable reproduction of this same dilemma for other cultures.

This chapter will therefore focus on the ethical implications of the idea of authenticity in tourism. First, it will examine the idea of authenticity expressed by tourists themselves, then examine the role of the tourism industry itself through an analysis of the ways that destinations are marketed and the manner in which local cultures are commodified and performed for external consumption. Lastly, it will look at nature as a source of authentic experiences and values and at some of the broader ethical implications for host and guest of the search for authenticity.

Tourism, ethics and authenticity

Although, as Urry (1991: 51) notes, '"the search for authenticity" is too simple a foundation for explaining [all of] contemporary tourism', it is still critical when it comes to explaining cultural, ethnic or historical tourism. This is, perhaps, especially the case in tourism in the South, where local landscapes, cultures and practices are presented in a particular way to appeal to Northern holidaying tastes. In many ways, those who travel to the South for vacation experiences are looking for the authentic, the real and the genuine. However, it is clear that what is presented as real and authentic is often a kind of 'staged authenticity' (MacCannell, 1999) which is created and performed by local people in conjunction with the tourism industry to fulfil the fantasies and desires of visitors.

The tourism industry is very much dependent on the creation and perpetuation of such fantasy ideals, and the production and consumption of tourism images sustains a multibillion-dollar global industry. Edwards (1996: 197–200) suggests that the long-haul travel industry is increasingly dependent on ethnographic imagery to sell its 'product'. Such images are created and presented as objects that can be observed and that have a verifiable, quantifiable authenticity, such as hand-woven rugs, local crafts, and re-enactments of dances, plays or 'everyday' labour. Urry (1994: 236–238) argues that tourism is all about the consumption of signs, images and texts, and therefore much of what is involved in examining the issue of authenticity in tourism is about an investigation of cultural interpretations of these signs – that is, of semiotics. The

> tourist is interested in everything as a sign of itself. . . . All over the world the unsung armies of semioticians, the tourists, are fanning out in search of the signs of Frenchness, typical Italian behaviour, exemplary oriental scenes, typical American thruways, traditional English pubs.
>
> (Culler in Urry, 1997: 3)

The authentic might, then, be defined as something that offers itself to the tourist as a 'sign of itself'. However, since the satisfaction of tourists' expectations depends upon consuming such signs, it is in both the tourist industry's and tourists' own interests to collude in their production.

Brown (1996: 33–36) suggests that tourism is filled with 'genuine fakes' that are the product of a collusion between the presenters of an attraction and those who visit it. Dann similarly argues that tourists play at reality in their pursuit of the kind of fun portrayed in the promotional material for particular destinations. In their search for the authentic they enter a kind of fantasy land where pseudo-events are staged and performed for their delight (Dann, 1996b: 117–125; for further discussion see Rose, 1999). Accordingly, the 'genuine fake' is not just the object itself – that is, the event or the artefact – but the relationship between the visitors and the presenters that the object mediates. In the tourist search for the authentic and the 'really real', the actual reality of the hosts' everyday lives – the actual landscape, and so on – becomes identified with the fake.

This can readily be seen in tourists' own construction and presentation of their holiday experiences through photography. Through their creation of a photo-graphic record, tourists exercise considerable power and control over the construction of the tangible memories of their tour. They produce a fantasy world, a perfect destination where they had a perfect time. Tourists can actively engage in producing idealized images in their photographs, with the collusion and assistance of local communities and tour operators. For some tourists it can be important to record what the local community presents as the 'best image' of their village, landscape or urban area. Markwell's (1997: 148–155) study of photog-raphy during a trip to Malaysia revealed that many of those on the tour had operated a selective filter to take photographs as a supposedly genuine and verifiable record of their authentic experiences. Markwell notes that many of the tour participants had taken a picture of the hut accommodation they were staying in but managed to frame the photograph in such a way that the satellite dish attached to the hut was cut out. This pictorial selectivity served to reinforce the myth of the perfect holiday in the perfect pre-modern world rather than demonstrating the problematics of travelling in the reality of a less than ideal world. Tourists themselves collude in the creation and perpetuation of an idealized and authenticated image of their chosen destination, one that excludes the inconveniently placed signifiers of modernity, of poverty, environmental degradation, or social and political decay. For Munt, the new middle-class tourists from Western societies have increasingly aestheticized and fetishized developing countries as a tourist destination, and glossed over the realities (Mowforth and Munt, 1998: 125–187; Munt, 1994a; 1994b: 49–60). In this sense, despite deploying a moral language of authenticity such tourists are very far indeed from the ethical ideal of respect for difference.

It is the search for the authentic that creates tourist motivations to visit new and 'unspoilt' peoples and places. It is clear that tourists, especially those who seek out holiday experiences in the South, search out the 'genuine' travel experience that might help them define their own self-identities over and against the exotic 'Other' they intend to visit (see Chapter 5). However, as Brown argues, the tour itself frequently reveals the attractions to be essentially inauthentic because they are timetabled and staged for the delight of external visitors (Brown, 1996: 36–39). Consequently, the tourism industry shapes and defines the authentic/inauthentic dichotomy because it exerts control over scheduling events and the holiday experiences. The tourist experience is highly structured, and through routeing, zoning and signposting, the tourism industry highlights attractions and tells tourists where and when they may gaze (Dann, 1996b: 77–79; Markwell, 1997: 138–141). This is even the case with independent and flexible holiday making, where tourists define themselves as travellers who exert control over the pace and direction of their 'more authentic' local experiences. But even these 'independent' travellers are reliant to some degree on travel literature and tour guiding services which in effect tell them where to go and what to see (Mowforth and Munt, 1998; Munt, 1994a).

In this way a collection of projected itineraries and images is built up, and these establish the boundaries of tourist experience. The images define what is beautiful, what is an attraction, what should be experienced and with whom one should interact (Dann, 1996a: 79; Pearce, 1995: 143–148). These images often *re-present* notions of the 'really local' in such a way that the ordinary and everyday can be captured and used to appeal to the tastes of Northern societies, so that basket weaving can suddenly be transformed into the exotic (Edwards, 1996: 207–211). The images presented by the tourism industry are often imbued with notions of the remote, the primitive, the unspoilt, and these are used as markers of tourist desirability.

Tourism brochures and guidebooks often carry pictures of landscapes such as deserted beaches that have no people in them at all in order to communicate the impression that this holiday will be a 'getting away from it all' experience (Dann,

1996a: 61–64). The tourism industry uses such images to reinforce a sense of placelessness and even timelessness (Dann, 1996b: 125). Nosy Be in Madagascar (Plate 6.1) provides an excellent example of this since it is 'sold' as an idyllic beach destination, with the added attraction of wilderness areas inhabited by the charismatic and often elusive lemur. Notions of wilderness are constructed and defined through a Western lens. In travel brochures and books, depictions and descriptions of unspoiled wilderness are clearly intended to whet Western appetites for what they regard as pristine and exotic – the essential pre-requisites for a 'getting away from it all' experience (for further discussion of wilderness, see Brockington, 2002; Wolmer 2002). The key feature of any wilderness is that it is devoid of people, and exudes the idea of the harmonies and natural rhythms of nature. Of course, very often those environments are locally regarded as working landscapes that are used for agriculture, the collection of plants and medicines, or for hunting, and so in reality they are often filled with people. One good example of this is the Pyramids and Sphinx at Giza. They are often photographed and depicted as though they are to be found in the middle of desert, in an area devoid of people. Yet in reality they are sited on the outskirts of Cairo and are surrounded by shops, offices and homes, not primitive, timeless and people-less desert.

The depiction of local cultures is equally important in terms of sustaining the image of local authenticity that underpins the long-haul travel industry. Tourists can be led to believe that what they are looking at is a timeless culture poised on the edge of change, hence the exhortations in much tourist literature to see it before it is too late. The 'Other' that is presented in the images used by the tourism

Plate 6.1   Idyllic beachscape, Nosy Be, Madagascar, 2001

industry is the antithesis of modern man or woman located in the Northern industrialized world. Instead, exotic cultures are often portrayed as mere extensions of the natural world, their inhabitants living close to and in harmony with nature. Numerous tourism brochures depict local cultures engaged in hunting, body painting, spiritual rituals and dancing. In this way the purity of wilderness and landscape merges with the idea of a purity of local culture in a mutually sustaining staged authenticity (Edwards, 1996: 200–211; Wilkinson, 1992: 386–389). The local people that appear in the brochures are often pictured in relaxed poses, scantily clad and constructed as sexually available (Dann, 1996a). They are used as part of the scenery, as objects rather than subjects, in order to appeal to the spectator tourist, and this too is clearly the opposite of an ethical relation that treats people as ends-in-themselves (see Chapter 4). In this way objects and people become interchangeable in tourism imagery so that the run of pictures in brochures will often present an animal, a landscape, a hotel and a local person to appeal to the Western definition of the ideal exotic location – although the images of local cultures clearly have to strike a balance between making sure that local cultures look exotic, unfamiliar and authentic, and ensuring they are regarded as safe, welcoming and attractive by the potential consumer. However exotically they are dressed, the local people of tourist brochures are always smiling or laughing.

These images of authenticity clearly have important ethical implications for the locals themselves. This emphasis on local exoticism can lead to inventions of traditions to satisfy external definitions of what is genuine (for further discussion, see Hobsbawm and Ranger, 1983). In turn, the people in the tourist destination may feel forced to adapt their lifestyles to ensure that tourists are not disappointed (Hall, 1994: 174–182; McKercher, 1993: 12–14). Craik (1995: 87–89) suggests that even new forms of tourism, such as ecotourism, adventure tourism and cultural tourism, have become increasingly intrusive and dependent on the destination community, using their culture as a resource. This, of course, is the very antithesis of an ethical relation, and such clashes also raise the issue of moral relativism (see Chapter 2). That said, the analyses which point to the detrimental effect of tourism on indigenous cultures often fail to address how local cultures can respond positively to contact with visitors. Local cultures often prove to be highly resilient and capable of interacting with tourists so that both hosts and guests return with some valuable experiences (Abbott-Cone, 1995: 314–327; Brown, 1992: 361–370). It is clear that the outcome of host–guest interactions is more complex than a simple case of 'guest exploits and dominates host'.

Image and desire in tourism: Belize

We can perhaps best elaborate on these complex inter-relations through a more detailed look at a specific destination. Tourism is a relatively recent development in Belize, and really began with the expansion of facilities during the 1980s. The tourism industry in Belize has three key attractions: rainforests, Mayan ruins and coral reefs – many visitors come specifically to dive on the second largest barrier reef in the world. Once Belize was dependent on primary commodities, but tourism is now its single largest foreign exchange earner, and the Belize Tourism Board (BTB) estimated that by 1994 tourism was earning the country B$150 million (US$75 million) (interview with Maria Vega, Vega Inn and BTIA, Caye Caulker, 21 January 1998; McField *et al.*, 1996: 97; *San Pedro Sun*, 1997c, d). The expansion of tourism has significantly changed the economies of the resort areas such as Caye Caulker and San Pedro, which were largely dependent on fishing (interviews with Dwight Neal, Director, Marine Research Centre, University of Belize (UCB), Calabash Caye, 24 January 1998; Mito Paz, Director, Green Reef, San Pedro, 2 February 1998; James Azueta, Marine Protected Areas Co-ordinator, Fisheries Department, Belize City, 9 February 1998; and Mike Fairweather, Resort Developer, Calabash Caye, 13 January 1998; *San Pedro Sun*, 1997b). As Plate 6.2 indicates, the growth of tourism in Belize has been backed by the government and by the private sector. The people of Belize have been directly encouraged through advertising to embrace the development of the tourism sector.

This rapid development in the tourism industry was partially dependent on selling an image of Caribbean beauty and impenetrable Central American rainforest to potential customers. When we interviewed tourists in Belize about their experiences it was clear that notions of the authentic and inauthentic were vitally important to them. In interviews, tourists indicated what they regarded as important sites/sights and genuine cultural experiences. Their comments revealed a desire to experience an exotic and unspoiled landscape, a real Garden of Eden. Tourists in Belize had a very clear idea of what they believed was authentically Belizean in terms of both the natural and cultural environment, and tours reflected

Plate 6.2 Sign urging local people to assist in developing the tourism industry, Belize City, 2000

this through their timetabling of reef, rainforest and ruins experiences for eager visitors. The image of Belize as vacation destination and the tourists' expectations were often informed by stereotypical representations of the Caribbean and Central America. For example, one tourist remarked, 'my image was pirates, seriously, and palm trees' (interview with Liam Huxley, San Pedro, 25 December 1997). A number of tourists stated that Belize had conformed to their expectations of the Caribbean as constituted by islands and sunshine (interviews with Jim, San Pedro, 3 February 1998; Shawn Nunnemaker, Caye Caulkar, 3 January 1998; David Sneider, Caye Caulker, 12 December 1997; and Fabrice Zottigen, San Pedro, 27 December 1997). Another part of the image is of the Caribbean as filled with reggae music and as a 'relaxed place to chill out' (interviews with Eddie D'Sa, San Pedro, 25 December 1997; Dave, Caye Caulker, 3 January 1998; and Yvonne Vickers, San Pedro, 30 November 1997) or as 'unhurried where tourists are forced to relax' (interview with Mary Tacey, San Pedro, 23 December 1997). It was notable that, in general, tourists wanted to see certain images, including palm trees, turquoise waters, and so on. Beaches (or lack of them) were mentioned frequently, and a recurrent theme was the desire to gaze upon exotic peoples, whom they identified as indigenous Mayans and Rastafarians (interviews with Shawn Nunnemaker, Caye Caulker, 3 January 1998; and Steve, San Pedro, 3 February 1998).[1]

The role of the industry itself was vitally important to any understanding of what tourists wanted to see and do as part of an authentic holiday experience. Interviews with tourists revealed the influence of tour operators in constructing a list of must-sees and must-dos for visitors. When diving or snorkelling, tourists were keen to see a list of spectacular marine life, rather than a general coral reef scene with numerous small fishes. This was noticeable in terms of tourist desires to see larger or brightly coloured marine life, such as lobsters, barracuda, sharks or parrotfish (interviews with Barbara Burke, San Pedro, 22 December 1997; Mito Paz, Director, Green Reef, San Pedro, 2 February 1998; and Dan Smathers, San Pedro, 3 December 1997).

The need to see large fish or wrecks is rather like the marine equivalent of the African Big Five on which the wildlife safari industry depends (for further discussion, see Barbier *et al.*, 1990; Brockington, 2002). Consequently, those involved in tourism and conservation in Belize have recognized the importance of allowing tourists easy access to spectacular marine life. The importance of Shark Ray Alley and Hol Chan Marine Reserve trips for snorkellers, and the Amigos Wreck dive or Blue Hole dive for divers, illustrates this. Shark Ray Alley and the Amigos Wreck are well known in San Pedro town (Ambergris Caye) because tourists are guaranteed a sighting of sharks and rays, and are often encouraged to swim with them, feed them and touch them. This was constructed and presented as an authentic encounter with genuine wild creatures. In fact, like all the wildlife, the sharks and rays had become accustomed to being fed. They had originally begun to gather at Shark Ray Alley because it was where local fisherman had gutted and cleaned their catch. It was only later that guides identified it as a potentially lucrative tourist site (interview with Melanie Paz, owner, Amigos del

Mar Dive Shop, San Pedro, 1 February 1998). The Amigos Wreck too was specifically sunk by the Amigos del Mar Dive Shop because tourists enquired whether there were any wreck dives in the vicinity of the island. The wreck was intended to satisfy tourist imaginings of sunken Spanish treasure ships laden with bullion from the New World and to reduce pressure on other popular but overcrowded dive sites (ibid.).

Tourists have a list of sights they want to see in Belize that have to be ticked off to complete the travel experience before they return home with new tales to share with their families and peer groups. One tourist remarked that before coming to the islands in Belize he had 'done the tubing in a river in San Ignacio, we saw Mayan ruins in Yucatán [Mexico] and we can tell you all about the pyramids at Tikal [Guatemala] and Caracol [Belize]' (interview with Fred, Caye Caulker, 15 December 1997). Similarly, another tourist stated that his trip to Belize was motivated by his wish to see a big cat: 'that was the reason for coming to Belize, I want to see a jaguar, it is that thing of ruins, rainforest and reef' (interview with Shawn Nunnemaker, Caye Caulker, 3 January 1998). It is significant that he used this same phrase – 'ruins, rainforest and reef' – because it is a marketing slogan used by the BTB and private tour operators to promote the country internationally as a tourism destination.

The 'authenticity' of the tourists' experience of Belize is thus linked to their ability to gaze upon or perform the requisite tour highlights. This is hardly an example of remaining ethically open to the unexpected differences we might encounter, since it seems to reduce the encounter with otherness to something both controlled and consumable. This also links us back to issues of self-discovery: finding a supposedly 'authentic' experience helps claims to found an 'authentic' self. For divers in Belize the trip to the Blue Hole for a dive can be an all-important experience. One interviewee suggested that she liked to look at the coral formations rather than the fish to experience the beauty of the marine world, which she found to be a very *spiritual* experience (interview with Barbara Burke, San Pedro, 22 December 1997). The language employed here certainly suggests a genuine openness to experiencing a different environment respectfully as something both desirable and wonderful, but sometimes the relationship is more prosaic, and a failure to complete the tour itinerary comes to be regarded as a lost life-opportunity or a lack of closure. One recently qualified diver expressed great disappointment that his inability to equalize (resulting in ear and sinus problems) meant that he had to abandon his attempt to dive the Blue Hole (interviews with Petra Barry, San Pedro, 4 December 1997; Bob Goodman, San Pedro, 5 December 1997; Peter Liska, San Pedro, 25 December 1997; Shawn Nunnemaker, Caye Caulker, 3 January 1998; and Dan Smathers, San Pedro, 3 December 1997).

Many tourists in Belize did express concerns about how tourism might be affecting the local culture. One tourist was concerned by the cultural impact of tourism on local society, and she expressed sadness about what she termed a 'bastardized' culture in Thailand as a result of the growth in tourism. She noted that as a result of such cultural changes it was not worth visiting Thailand any more because it would not be an authentic experience of a different culture (interviews with Mindy Franklin, San Pedro, 1 December 1997; George Mackenzie and Katy Barratt, San Pedro, 2 December 1997; and Simon, Caye Caulker, 19 January 1998). This kind of remark is reminiscent in many ways of the complaints of earlier travellers like Ruskin that European travel was becoming less authentic (see Chapter 2). But here too it is often difficult to distinguish between the ethical labour which people like Ruskin felt to be inherent to authentic travel experiences and what Simmel regarded as egoistic enjoyment masquerading as a search for educational and moral values. For example, another diver remarked, 'I like wreck diving and deep diving, I have seen enough fish now. I like the adrenaline really' (interview with Tony, Caye Caulker, 12 December 1997). There is little evidence in this statement of ethical regard for the Other, in this case the marine environment, which has become something that facilitates and is subservient to an egoistic adrenaline rush. These complexities of unpacking the relationship between tourist/travel destination and self-discovery are brought out in Birkeland's (2002) account of modern travellers' tales of visits to North Cape in Norway.

It is true that the search for the exotic and genuine other is especially important for tourists who choose more independent holidays or who go to less easily accessible destinations in the developing world. The choice of small local hotels, restaurants and guiding services is all intended to satisfy the idea of what is authentic, while structured, pre-booked tours are defined as inauthentic and staged. The people who visit Belize are particularly interested in unstructured travel. It was noticeable that a number of those interviewed had previous experience of travelling in developing countries. It was clear that they had been backpackers in the past as students and were keen to continue with their holiday experiences in the same spirit. Their choice of destination and form of travel also reflected the fact that the majority were middle to high income earners, and they had chosen independent and unstructured backpacker-style trips with more comfortable accommodation (interview with Eddie D'Sa, San Pedro, 25 December 1997; see also Elsrud, 2001; Hampton, 1998; Murphy, 2001). Such tourists are much more likely to prefer vacation destinations that are perceived in their home societies as different or unusual, impressive, adventuresome and exciting (Lee and Crompton, 1992: 732–737). For example, one tourist mentioned that he looked at Belize on a map and thought it would be an exciting place, 'like you could get malaria there' (interview with Leia and George, Caye Caulker, 2 January 1998). Is this indicative of a search for moral authenticity, or simply an egoistic attempt to impress one's peers? Some tourists clearly go out of their way to endure difficulties to attain a more 'authentic' relation with their destination aim, but they also gain 'cultural capital' from retelling these stories. In this sense, 'authenticity' is always linked to issues of cultural distinction, of setting oneself apart as someone with a more genuine travel experience (Bourdieu, 1998). Such 'hardships' are in any case only ever relative, since even these more rugged forms of travel are still highly organized by tour companies either at home or in the destination, and so backpackers tend to be insulated from the more uncomfortable realities of life and travel in developing societies.

It is clearly difficult to separate out the real and the fake, the staged and the genuine, since all these themes are intertwined in tourism activities, and this difficulty is linked to difficulties in assessing what constitutes a genuine ethical relation. Tourism sights and activities are, at the same time, real and unreal. What seems undeniable, though, is that the definitions of the 'really real' that tourists clung to were partially informed by the images presented by the tourism industry itself.

Destination image and the politics of authenticity

The ways in which tourists define the authentic as distinct from the inauthentic, the genuine from the staged, are informed by their engagement with the global tourism industry. The private sector, governments and non-governmental organizations (NGOs) all assist in the creation and promotion of a specific image for consumption by tourists. This social construction and presentation of a destination is partly a response to existing consumer expectations. However, what is clear is that there is a mutual feedback between what tourists want or expect and what the tourism industry deems to be a suitable set of sights and experiences to satisfy their consumer base. For example, one tourist expressed her desire to see Belize City redesigned and rebuilt to fit tourist images of a cute colonial town with quaint inns and shops that would enhance its attractiveness to visitors (interview with Brie Thumm, San Pedro, 30 November 1997; see also Pattullo, 1996: 63–71).

This is a powerful indicator of the ways in which tourists want to see a particular kind of heritage. Some visitors to West Africa specifically go there to see places associated with the transatlantic slave trade, and to understand more about its history. However, the heritage and tourism industry is often engaged in presenting a particular non-threatening, non-political version of history. This version appeals to images of 'quaint colonialism' rather than a more political history of colonial rule. For example, hotels and restaurants can often make reference to the colonial past in developing countries. In the Caribbean, one of the most common images that draws on the past is the use of sites of past slavery in the form of old plantation houses renovated to perform as hotels and tourist attractions (Harrison, 1992c: 19–22). Here, then, the question of authenticity and ethics is one of temporality as much as spatiality, of seeing the past as it 'really' was. This ethical relation to the past is no less problematic (Smith, 2001b), because it becomes a question of whose history is being presented and for what purposes.

The construction of an 'authentic' destination image can assist governments in developing countries in the process of nation building and identity formation. In particular, the marketing strategies in developing countries often emphasize images of environmentally unspoilt landscapes, political stability and culturally interesting local people. Such images can be used for internal political advantage. Lanfant (1995) argues that tourism marketing shapes the image of a place, and the identity of a society is described according to seductive attributes and crystallized in a publicity image in which the indigenous population is insidiously induced to recognize itself. The state can then exploit a tourist image that flatters national identity and praises the nation-state in order to reinforce national cohesion. For those involved in tourism, the projection of a state as a tourist destination creates an idea of the state itself, which can in turn be fed back into the domestic political order (Lanfant, 1995; McCrone *et al.*, 1995: 12–17). In this sense, then, tourism, and the images of authentically national landscape/culture it employs, are party to the creation of what Anderson (1991) refers to as an 'imagined community'.

However, alongside this particular definition of national identity for marketing purposes, tourism also relies on using regional identities and sub-regional identities to develop the industry and increase profits. The regional, cultural and political identities used by the tourism industry also serve to define images of what it is to be genuinely Caribbean, Latin American, African or Asian. Likewise, sub-regional or sub-national identities are often used by the tourism industry. For example, Bali uses its particular geographical features as an island and its distinct Balinese culture to market itself as separate from Indonesia (Harrison, 1992b). Similarly, in Kenya the separate Maasai ethnic identity is used to sell the country to an external audience (see Hitchcock *et al.*, 1993; Sindiga, 1999).

McCrone *et al.* (1995: 8–12) argue that the commercialization of culture also entails an *ideological* framing of history, nature and tradition. Tourism can re-define social realities, so that where advertising creates images of a place, these imaginings then create expectations on the part of the visitor, which in turn lead the destination to adapt to such expectations. Since destinations have to compete with each other in the international tourism market, their histories, cultures and environments become subject to endless reinvention within the confines of a tourist gaze from which they cannot escape without abandoning their status as a destination. Such representations of cultures for tourism have vital implications for collective and individual identities within the destination country (Hall, 1994: 178–182; Urry, 1990).

Heritage is, then, both an important aspect of destination identity and a politically contested term. As such the narratives of heritage often reveal the social, economic and political relations implicated in their production (see Ateljevic and Doorne, 2002). For example, the Matopos National Park in Zimbabwe has a contested history that reveals different narratives of the land-scapes that privilege white over black history. Tourists largely visit the park to see symbols of white heritage, typified by Rhodes's grave and the monument to the Jamieson Raid. However, the park is also the site of important symbols of black history in Zimbabwe, most notably the grave of Chief Mzilikazi. While tourists generally do not see the sites that are important to Ndebele political heritage, they are taken to see ancient San bushmen paintings (see Alexander *et al.*, 2000; Ranger, 1999).

The questions surrounding *whose heritage* is presented to visitors are ethically as well as politically significant. Taylor argues that the presentation of Maori culture in New Zealand is largely dependent on the essentialization of 'Otherness' that is negative (Taylor, 2001). Similarly, the presentation of the Rocks area in Sydney is imbued with the rhetoric of Australian nationalism. This version of heritage has effectively silenced and suppressed narratives that highlight oppression, racism and conflict as part of Australian history (Waitt, 2000; see also Aitchison, 1999; Robb, 1998; Van den Berghe and Flores Ochoa, 2000).

The subtle (and sometimes not so subtle) interplay between authenticity and politics is also evident in the way in which Belize has become involved in developing itself as part of a broader regional identity to attract potential tourists. The debate about exactly whose heritage is being presented is critical in Central America. Brochures, books and guides often airbrush out the complexities of current Mayan politics in Central America in favour of a more romanticized view of Mayan culture in the past rather than the present. On a regional level, Belize is part of the Mundo Maya marketing strategy. Mundo Maya is an organization that covers states that were once sites of the Mayan civilization, namely Mexico, Belize, Guatemala, Honduras and El Salvador (Chant, 1992: 85-88). Mundo Maya markets tourism routes through the region based on the idea of travelling between archaeological sites such as Chichen Itza (Mexico), Tikal (Guatemala), Caracol (Belize) and Copan (Honduras) (interview with Pat Wiezsman, Executive Secretary, Mundo Maya, Belmopan, 27 November 1998). The agreement between the countries involved in Mundo Maya is intended to facilitate the flow of tourists (and therefore the revenue derived from tourism) so that they can travel more freely through the region. One aspect of this is that the Mundo Maya organization has lobbied for co-operative agreements between regional airlines that would, for example, allow Mexican airlines to enter Belize to drop off and pick up passengers en route to a fellow Mundo Maya country (*San Pedro Sun*, 1997e).

In this sense, depending upon the political context, tourist developments can help overcome as well as establish rigid nationalistic stereotypes, but the creation of a regional identity for tourism marketing purposes is never unproblematic. One of the difficulties associated with such regional plans is that some partners stand to gain more than others. For example, Belizean government officials refused to accede to a request from the Mexican government that the Belize barrier reef be renamed and marketed as the Maya Reef or El Gran Arrecife Maya. The Belize barrier reef is part of a much larger reef system that stretches from Honduras in the south to Mexico in the north. While the reef is marketed as part of the Mundo Maya experience, the Belize government was concerned that Mexico had already degraded many of its reefs (especially around Cancún) and so the Mexican tourism industry would benefit disproportionately from claiming that the Maya Reef was in Mexico. In effect the Mexican tourism industry would make financial gains from giving the impression that less environmentally damaged reefs in Belize were within Mexican borders (*Amandala*, 1997c). It was contended that potential clients would book their holidays in Mexico, as the better-known destination, and only discover on arrival that the photos of reefs used to entice them were taken in Belize.

The identification of Belize with Central America has also raised questions about how Belize fits into the perception that there is large-scale political violence in the wider region. Brochures aimed at potential clients from the North are one area where images of safety and security are critical in helping to determine travel decisions. For example, one brochure from UK-based Reef and Rainforest Tours draws attention to the relative stability of Belize and its distinct culture that sets it apart from the countries it shares borders with. The brochure states that 'Belize is an oasis of peace in the often-volatile region of Central America. Indeed, it sometimes seems more like a Caribbean than a Central American country' (Reef and Rainforest Tours Brochure, 1997). As a result, Belize has tried to distinguish itself from the political violence associated with the rest of Central America. Certain aspects of law enforcement (such as the Tourism Police) have been advertised and emphasized in order to create an image of safety and stability. Sometimes, however, the requirement for an orderly society has led to the use of coercive measures against local populations in order to make the destination appear more appealing to Western tourists (Matthews and Richter, 1991: 125–128; Stonich *et al.*, 1995: 21–24). Here the rights (see Chapter 4) of local peoples are often overlooked, and narratives that focus on the treatment of indigenous peoples by governments, the private sector and NGOs or on political strife and violence are effectively silenced.

The use of local cultures and the ways that less palatable aspects are suppressed in tourism marketing have also arisen in arguments over the ways that indigenous communities have been used as an attraction. The way that Mayan culture has been commoditized to sell the region as a tourist destination has not been without its critics, which include Mayan organizations. The growing interest in the potential benefits of community conservation and community-based tourism has also raised concerns about the ways that local people and their ways of life are customized, packaged and sold for consumption by foreign tourists (Butler and Hinch, 1996; Mann, 2000; see also Chapter 7). The most obvious example of this is where traditional rituals and festivals are re-enacted for the tourists' gaze. In addition, the tourist intrusion has brought social and cultural change with it that is more in line with commercial values (Hall, 1994: 130–133; Pattullo, 1996: 84–90). However, critics of this position argue that this theory exaggerates local susceptibility to tourist lifestyles. Harrison points out that while tourism has been accused of degrading the meaning of local rituals, they may not lose their impact and importance for the local people performing them. Local people may not simply be staging a 'primitive culture' to appeal to Western tastes and definitions of other-ness. Instead, local people may use those performances for their own social, political and cultural purposes. As we have seen, in Bali the performance of Balinese rituals for tourist consumption has been used as a means of reasserting national identity in response to the increasing threat of Indonesianization (Harrison, 1992c: 19–22; see also Hitchcock *et al.*, 1993).

Mayan villages constitute a major cultural attraction for international visitors. International tourists particularly seek after traditional Mayan crafts, and Mayan communities have actively responded by developing wood and stone carving, basket weaving and textile production. One of the difficulties with this response is that Mayan crafts are then divorced from their cultural and religious context, and thereby lose a great deal of their significance. For example, tourists buy and use the brightly coloured woven bags (*cuxtal*) that are traditionally used by Mayan men when working in the *milpas*. Likewise, the complex patterns that are woven into women's shawls (*huipiles*) tell a story of spiritual and historical significance to the onlooker, so that different colours in the shawls might represent various spiritual forces and environmental elements. The *huipiles* that are crafted for international tourists end up as art pieces and wall hangings in homes in North America and Europe (Abbott-Cone, 1995; Cohen, 2001; Moreno and Littrell, 2001; Toledo Maya Cultural Council and Toledo Alcaldes Association, 1997: 26).

Healy (1994) points out that for tourists, authenticity is very important in the objects that they buy on holiday. The knowledge that something is handmade is often critical in determining whether a tourist defines the object as authentic or not. As a result, it is not uncommon to see craft workers producing goods such as baskets in front of tourists. Healy suggests that the production of handicrafts from local materials provides the poorest of the poor with an opportunity to engage with the tourism economy and derive some benefit from it. Not only that, but the production of souvenirs based on traditional objects may actually revitalize craft traditions (Healy, 1994: 143–149; see also Bunn, 2000: 166–193; Stiles, 1994: 106–111; Teague, 2000: 194–208). A similar example is the reinvigoration of traditional dances and spiritual rituals that villages no longer practise for a variety of reasons. Conejo Creek in the Sarstoon-Temash area has re-established its traditional deer dance with funding from the Kekchi Council of Belize. This funding allowed dancers to rent costumes from Mayan communities in Guatemala that still practised the deer dance. However, Mayan leaders have pointed out that these rituals and traditions should be revitalized to ensure that Mayan culture will continue, and not simply for tourist consumption (interview with Gregory Ch'oc, Kekchi Council of Belize, Punta Gorda, 23 May 2000).

While the production of local crafts for consumption by tourists can often bring significant benefits to the local populace, and might therefore seem to be favoured by a utilitarian approach (see Chapter 3), it is not unproblematic. In Belize preparation of *jipijapa* and *bayal* plant fibres for the baskets requires a great deal of time and labour, and then skilled craftspeople are needed to weave the baskets (Healy, 1994: 137–151; Toledo Maya Cultural Council and Toledo Alcaldes Association, 1997: 28). Consequently, the baskets are expensive compared with other souvenirs, and tourists then complain that since the baskets are only plain and unpainted they are not worth the price. Of course, local people who attach a cultural and spiritual value to the baskets view this as an insult. Traditionally the baskets are special because they are used on particular occasions and have a spiritual dimension. Yet they are in danger of becoming merely trinkets that have to compete with other arts and crafts in the markets offering tourist merchandise (Saqui, 2000). According to Pio Saqui at the University College of Belize, it is not uncommon to find Mayan children who believe that the baskets are made purely for tourists and have no other significance beyond their capacity to make money for their families. Bramwell and Lane (1993: 75–79) argue that there is always the underlying danger that culture, including rituals, local products and festivals, will become a commodified heritage asset that lacks any moral meaning for those expected to perform it for the tourist gaze.

All this goes to show that the ways in which local history is constructed and communicated can be highly political. In the case of the Mayan communities in Central America, their history of persecution at the hands of governments in the region is often left out of any presentation of Mayan culture to external visitors. Rather than benefiting from the use of the name Maya as a brand, Mayan communities in Belize have felt that their temples, rituals, way of life, history, culture and people are merchandised to raise revenue for private businesses (specifically tourism) and governments without providing any particular benefit for them. Instead, Mayan communities have the highest rates of infant mortality, illiteracy, poverty and malnutrition in Belize (see Chapter 7) (*Amandala*, 1998, 1999f; Saqui, 2000). This stands in stark contrast to international marketing of Maya culture by the government of Belize and other Central American states to attract international tourists and the revenue that they bring with them. In tourism marketing literature, Mayans are constructed and presented as a people untouched by modernity and living a simple, agrarian way of life, wearing traditional clothing and engaging in age-old spiritual rituals. In reality, the relationship between Mayan communities and modernity is significantly more complex and definitely different from the image presented to international tourists. For the majority of Mayan people, theirs is an experience of social, political and economic marginalization, and even exclusion.

Politicians, diplomats, lending agencies and consultants have all proclaimed tourism to be an engine of growth, and its images of power and prosperity have been touted to launch areas like the Caribbean into development and modernism and out of poverty (Pattullo, 1996: 2–5). Yet ironically, the push towards modernization is based upon the presentation and appeal of the authentically traditional rather than the new, the natural rather than the artificial, and the past rather than the present, all of which is the very antithesis of modernity. Harrison argues that tourism is where the modern and the traditional stand in stark contrast to one another – a contrast most obviously seen in the effect of tourism on the physical landscape with the appearance of new buildings and amenities to serve the tourist populace. Yet traditional and modern, natural and cultural, past and present are constantly interwoven.

Nature and authenticity

As we have seen, nature has a particular role to play as a source of authentic experience. Tour operators in Belize use a variety of images that operate on three levels to appeal to potential travellers. On one level, tourism operators use green terms to describe their companies. For example, they are often presented as companies that assist First World tourists in exploring developing countries, as destinations with limitless natural resources that are simply waiting to be discovered by travellers. On another level, tourism operators utilize the language of sustainable development to legitimize their travel package. Finally, tours rely on the underlying themes of nature, nostalgia and nirvana. In these tours developing countries are presented as unspoilt. For example, national parks are presented as pristine wildernesses with no discussion of who established them and why or how such reserves affect local people (Brockington, 2002).

In 1994 Belize's then Minister of Tourism and the Environment, Henry Young, stated that for too long the derogatory expression 'banana republics' had been used to describe countries such as Belize that were seen as having too much jungle and too little development. However, for Henry Young that image of the corrupt and civil war-stricken banana republic was rapidly being replaced. The new image was one of a more positive perception of Central American states as places of rainforests teeming with biodiversity, filled with ancient cultures and offering a perfect place for visitors with an interest in, for example, the natural healing remedies of peoples living in the rainforest. So Northern tourists were seeking the Southern countries precisely because of their lack of Western-style development – that is, for precisely the same reason for which they had once been shunned (Young, 1994: 4–6).

The ways that Belize's environment is made to perform a role as a source of authentic tourist experiences is evident throughout the industry. The BTB's own promotional literature uses the slogan 'Friendly and Unspoilt' alongside photographs of the country's key attractions: its jaguars, Mayan ruins, rainforests, marine life and the Blue Hole (in a promotional pack for tourism distributed in 1997). A number of coastal resorts market themselves through the use of images of the Caribbean designed to appeal to Western tourists keen to relax among silver sand and turquoise water. For example, the Manta Resort, situated on the relatively underdeveloped Glovers Reef Atoll, claimed to have 'white sands, warm waters, pristine diving . . . no phones, no TV, no hassles',[2] while Blackbird Caye's promotional literature described it as 4,000 acres [1,600 ha] of remote, unspoiled tropical jungle, a virgin paradise where trails allowed visitors to explore the breathtaking beauty of the island (in a Blackbird Caye resort brochure from 1996). The notion of a pristine paradise designed to portray resorts as fantasy islands where the tourist can return to some kind of non-Western pre-modern primitiveness is commonplace in adverts that aim to attract tourists. In tour operator brochures Belize was also described as a place where

> denizens of the deep glide gracefully through pristine waters. Belize is *natural* and *unspoiled*. It's like a real-life Jurassic Park . . . *travel back in time*; visit villages as they were years ago; talk with villagers whose ancestors inhabited these lands before Christ was born.
>
> (Reef and Rainforest World Wide Adventure Travel, 1999)

A reference to films like *Jurassic Park* is clearly intended to spark the imagination of potential visitors, encouraging them to perceive Belize as a primeval rainforest area fortunately devoid of Western-style urbanization and development. But this image, like that of other cultures, has been carefully crafted. Not all aspects of nature are equally appealing. For example, the exotic island fantasies of potential clients persuaded one resort developer, Mike Fairweather (interviewed in Calabash Caye on 13 January 1998), to change the name of one of

Belize's coral islands from the rather unappealing Cockroach Caye to the rather more exotic Seahawk Caye to give it a more attractive image.

The extensive manner in which cultural ideals inform and alter our notions of nature and the value we place on it have led some to argue that wilderness itself is, at least in part, 'socially constructed' (Eder, 1996). It might almost be defined as a 'state of mind' (Nash, 1982: 5) rather than an authentic underlying reality. Although the idea of wilderness as authentic pre-human natural environment has played a large part in modernist and counter-modernist narratives, as something either to overcome or to preserve, we must recognize that it too has a human history. Nature itself is a culturally and politically contested category (Macnaghten and Urry, 1998). There is no such thing as pure nature, and while Belize has some of the most important and intact rainforests and reefs in Central and South America, none of the areas can accurately be described as wilderness if by this we mean that they are uninhabited or entirely undeveloped. (Indeed, many areas have been or are about to become the victims of massive development projects (Worrall, 2002).) In this sense too, the tourist brochure can lead to a stereotypical and essentialized view of the absolute otherness of a supposedly authentic nature rather than a more attenuated understanding of its different articulations with local cultures.

This is not to say that wilderness is just an invention; it is not (Oelschlaeger, 1991). Nor is it to say, as some theorists have argued, that nature can be of no 'intrinsic' ethical value, that it cannot be regarded as an end in itself. After all, the fact that we cannot have a truly authentic experience of another culture, one that is pure and unaffected by our own cultural expectations, does not rule out the possibility of an ethical tourism based on respect for that culture. Similarly, the fact that we cannot escape to some 'untouched' wilderness, and that our ideas of what counts as wilderness are affected by many cultural factors, does not mean that an environmental ethics is an impossibility (Smith, 2001a). Ethics is, as we have argued, all about socially mediated relationships and evaluations, and this is no less the case when it comes to nature. But there seems no doubt that the land and seascapes of countries like Belize can inspire the desire and wonder that theorists like Levinas and Irigaray regard as the source of ethical relations (see Chapter 5).

The ethics of authenticity

As the example of Belize shows, the issue of authenticity is complex and impacts on many different aspect of ethics and tourism development. There are at least two aspects to authenticity that can combine in the tourist experience to create what the tourist will consider to be a 'genuine' holiday experience, namely, judgements about the authenticity of the toured objects (baskets, rainforest, traditional dances, and so on) and the felt reality of the tourist experiences themselves. 'An authentic experience is one in which individuals feel themselves to be in touch with both a real world and with their real selves' (Handler and Saxton, 1988: 243). Yet as Wang points out, it would be mistaken to believe that there is an objective or simple causal relation between these two aspects:

It would be wrong to think that the *emotional* experience of the 'real' self ('hot authenticity') necessarily entails, coincides with, or results from the epistemological experience of a 'real' world out there ('cool authenticity'). . . . Certain toured objects, such as nature, are in a strict sense irrelevant to authenticity in MacCannell's [objective] sense. However nature tourism is surely one of the major ways of experiencing a 'real' self [what Wang refers to as existential authenticity].

(1999: 351)

Thus tourists may claim to have had an authentic diving experience even though they really know that the wreck presented to them is in some senses an artefact of the tourist industry, that they are diving in a marine preserve, and so on.

While the issue of authenticity pervades the discourse of heritage, cultural and nature tourism, there is no simple, objective way of defining what is real, traditional or natural. In so far as authenticity is defined in terms of the existence of alternative (non-modern) forms of life, or of untouched wilderness, then the very presence of the tourist makes such experiences problematic. What is more, this may not matter to tourists, who often actively collude in staging and performing authenticity (Edensor, 2000). So far as tourists are concerned, the ethical issue is not simply one of whether they are being conned by a specially concocted ceremony or fooled by an artificial landscape that is entirely 'fake'. These 'realities' are all negotiated, and all are dependent upon images and expectations developed through diverse media. Some of these images may be consciously constructed for their own ends by governments or the tourist industry; some are more peripheral, such as watching nature programmes on television or reading the *National Geographic* in the dentist's waiting room. The fact that authenticity is negotiated is not, however, to say that the tourist enters a (post-modern) social space where reality and illusion are indistinguishable. Clearly, there are limits to the degree to which 'reality' can be negotiated, but these limits will vary for different social groups with different agendas. Authenticity becomes an issue for tourists only when their expectations are unmet or their desires unfulfilled.

The issue of authenticity is much more important for the host community. This creation of authenticity in tourism has direct impacts on those people and communities who act as spectacles for consumption. Local people are insidiously encouraged to recognize and perform a caricature of themselves, which they play out for visitors. This caricature is imbued with Western definitions of what is deemed exotic, ideal or traditional. As a result, the mundane and ordinary tasks of local communities become redefined as exotic spectacles. In this way, traditional or genuinely local practices can be reworked to satisfy the needs of external visitors, so that what is really real can be invented and performed to conform to the idealized holiday fantasies of Northern tourists. The 'authentic' version that emerges can often silence alternative versions, especially those that might highlight oppression, racism or conflict (Waitt, 2000).

However, this is not to suggest that local cultures are always weak and unable to cope with their engagement with the outside world. The ways that Mayan communities have criticized the use of Mayan culture for the tourism industry indicates that they can be robust. Furthermore, local cultures can use the tourism industry's interest in traditions and rituals to inject new life into waning local enthusiasm. Some local cultures have managed to return the tourist gaze. Taylor argues that communally organized Maori Heritage Tours used their staged events and the signing of a Visitors' Treaty to get a message across to the tourist about the original Treaty of Waitangi. Tourists are guaranteed a 'vibrant cultural experience' and simultaneously encouraged to recognize 'their own position as consumers of culture' (Taylor, 2001: 21). While highlighting the 'inauthentic' nature of the staged act, this is nonetheless, Taylor argues, an (ethical) relationship based on 'sincerity'. (Perhaps this kind of relation could be understood in terms of a communicative ethics (see Chapter 5).) These kinds of examples indicate that notions of the authentic and inauthentic or the traditional and the modern are complex and interrelated.

Lastly, perhaps we need to recognize that to a large extent, tourists do not usually want to experience a genuine *ethical* relation with the Other, in Levinas's terms (see Chapter 5). They may experience both desire and wonder at the Other's existence, but as tourists their interest in sustaining the Other in all of their difference is at best marginal. (Which is not to say that the tourist relation may not blossom into something more: a real desire to preserve rainforests, to assist the poor of the South, and so on.) Even those tourists seeking to give depth to their 'inauthentic' modern existence by experiencing the exotic do so vicariously, at a remove from the local reality, through a commodified (instrumental) relation, and for a strictly limited period (a fortnight or month 'abroad'). The exotic they seek is often just a counterpoint to their own desires and expectations – that is, in Levinas's terms, it is reduced to the Same.

7   Ethics and sustainable tourism

Previous chapters have presented many examples of the kinds of ethical problems that can arise through tourism developments. We have framed these issues in terms of particular ethical approaches, including moral relativism, utilitarianism, rights, distributive justice, communicative ethics, and the ethics of care, difference and of authenticity. As we have seen, recent debates about the negative impacts of tourism have led to calls for a more environmentally aware and culturally acceptable form of tourism. In particular, the concern was raised about mass tourism that it was culturally insensitive and especially damaging to indigenous communities. This last chapter will chart responses within the tourism industry, which have sought to allay ethical concerns by addressing at least some of the issues surrounding the past exploitation of local communities and environments. These various forms of alternative tourism (Krippendorf, 1982), which include among others 'sustainable tourism', 'ecotourism', 'community-based tourism' and 'ethical tourism', generally involve small-scale developments that substitute for, or place limitations on, the unethical excesses commonly associated with mass tourism.

It should be said at the outset that there is considerable scepticism about the ability of alternative tourism actually to replace mass tourism (Cohen, 1987), given the latter's immense economic clout and its popularity with tourists themselves. The sheer numbers involved are staggering: 5,000,000 visitors per year are attracted to Benidorm's high-rise apartments, consuming some 12,000,000 cubic metres of water and producing 176 metric tonnes of rubbish per day. An area where 350,000 people squeeze into 30 square kilometres is clearly unsustainable in environmental terms, but Benidorm is still booming economically, as the opening of the new £60,000,000, 48-storey Gran Bali hotel shows (Tremlett, 2002). Alternative forms of tourism have been regarded as a means of reducing the impact of the industry. However, it is not always readily apparent that they do so. Even relatively small-scale tourism can have serious negative impacts (see Plate 7.1).

While alternative tourism may only ever account for a small percentage of tourist visits, the provision of more sustainable alternatives may serve to indicate other, more ethical possibilities for the industry as a whole. It may also provide alternative paths of development for particular countries and regions that do not wish to, or simply cannot, follow Benidorm's example.

Plate 7.1 Rubbish created by tourism in Placencia being burnt at the rear of the main village. Originally, tourism was only allowed to expand subject to an agreement that rubbish would be removed and disposed of on the mainland of Belize.

Following a general account of sustainable tourism, the first section of this chapter will pick up on the discussion of tourism in Belize by focusing on small-scale ecotourism to Mayan communities in Central America. The second section addresses some of the complex ethical issues arising out of what is probably the most quoted case of sustainable and community-based ecotourism, that of Zimbabwe's 'Communal Areas Management Plan for Indigenous Resources' (CAMPFIRE). These two projects have quite different relations to the natural environments that form a key part of their touristic appeal. Much of the income of the CAMPFIRE project is derived from sport hunting, whereas tourism to indigenous communities in Central America does not generally rely on consumptive use of the environment. This raises some of the questions we began to address in Chapter 6, namely, the possibility of widening ethical relations to include non-human members of a locality in terms of animal 'rights' and/or environmental ethics.

Tourism and sustainable development

The question of whether environmental care and development are compatible has been a focus of debates on sustainable development in the South. It is often argued that economic development, export-oriented growth and industrialization are forms of development inimical to environmental conservation. This stems from the belief that the natural environment must be exploited for development, and that this will probably lead to the kind of degradation and environmental changes previously witnessed by the industrialized world (see Hardin, 1968: 1243–1248; Meadows, 1972). The United Nations Conference on Environment and Development (UNCED) held in Rio de Janeiro in 1992 represented divided views on environment and development, and these views were again reflected in discussions during the 2002 World Summit on Sustainable Development in Johannesburg. Leaders from Southern states argued that environmental sustainability was a luxury they could not afford since it constituted an impediment to development. Industrialized states, on the other hand, were keen to point out that industrialization and development in the South constitute a global environmental threat. As a result, countries of the South were to be persuaded to curb industrialization in favour of the global environmental good. The South perceived the Northern position as one of double standards, claiming that the North contributed to most of the current global environmental problems and was least willing to provide finance to tackle them.

Current patterns of development that emphasize economic growth at the expense of social and political development have certainly damaged the environment. Whereas in the North environmental concern centres around a clean-up of the adverse by-products of industrialization, in the South environmental degradation is very much a survival issue. For countries that rely on exporting crops and other raw materials, degradation of the natural resource base can affect crop yields and, consequently, national income. At the other end of the political spectrum, environmental damage affects the rural dweller reliant on water, land and wood. A number of commentators have viewed poverty, inequality and environmental degradation as a self-reinforcing triangle (Bruntland, 1987; Chambers, 1983; Johnson and Nurick, 1995; Redclift, 1992). However, Woodhouse (1992: 97–116) suggests that for the industrialized nations, there are two dimensions of concern for sustainable development in the South: the fear that industrialization will increase global pollution and contribute to global climate change, and also the 'Third World' element to global problems, such as population growth.

Tourism has been increasingly promoted as a path to development that can satisfy the needs of environmental conservation and development. Tourism is often regarded as a way out of the classic problem of earning foreign exchange without destroying the environmental 'resource' base and compromising sustainability. Tour operators have not been slow to recognize the potential for marketing developing countries as sustainable tourism or ecotourism destinations (Cater, 1994: 69–72; Weaver, 1994: 159–162). Development strategies based on tourism are very much part of the neoliberal debate about comparative advantage, which advocates that each state should concentrate on exporting goods that it is naturally best at producing (Amsden, 1990: 5–32; Porter, 1990). Developing countries are considered to have a comparative advantage in tourism because they attract tourists from the North who seek sunshine, beaches and other natural and cultural attractions found in the South. Governments and their tourism boards, along with private enterprise, therefore project an image of a destination, and particularly of its natural environment, as something there to be discovered and enjoyed by tourists from industrialized nations (see Chapter 6). In essence, the emphasis on tourism is designed to increase economic diversification away from a few traditional exports, such as bananas in the case of Belize or tobacco in Zimbabwe (Brohman, 1996: 51–55; Pattullo, 1996: 11–13; Stonich *et al.*, 1995: 6–9). The commitment to community development through ecotourism thus combines ideas of sustainable development and 'green' capitalism, where natural resources are viewed as a means of generating revenue.

For these reasons, tourism has often been touted as a panacea for North and South alike. Even the motto of the World Tourism Organization – 'Tourism: Passport to Peace' – suggests that it increases communication and understanding through cultural exchange. However, critics of tourism development have pointed to its obvious shortcomings in the South. More tourism development does not automatically mean greater well-being (utility), and the operation of tourism has frequently failed to match the slogan (Hall, 1994: 89–91). Critics suggest that one of the major problems associated with tourism development in the South is that it exacerbates existing and creates new economic and social divisions in the host communities. In particular, enclave tourism has tended to create spatial inequalities through the establishment of all-inclusive resorts. These resorts cater only for tourists, and the conditions in the resorts starkly contrast with the living standards of the neighbouring local communities (Brohman, 1996: 55–59; Dieke, 1993; Lea, 1993: 707–710; Lindberg *et al.*, 1996: 547–551; Lynn, 1992: 371–377; Wallace and Pierce, 1996: 843–846). Consequently, it is vitally important that those involved in developing tourism pay special attention to the ways that tourism can benefit the host population (Krippendorf, 1987: 115–119).

One response to criticisms of the socially and economically divisive effects of mass tourism has been the development of community-managed tourism in the 'developing' world. Social or community tourism can be defined as tourism resulting from participation by economically weak or otherwise disadvantaged people with the intention of extending any benefits derived to these economically marginal groups (Bottrill, 1995: 45–48; Butler and Hinch, 1996; Hall, 1994: 43–47; Price, 1996). Community-based ecotourism is intended to maximize the participation of local people in decision making from a very early stage. In particular, local communities are expected to play a major role in deciding on the direction and rates of ecotourism development in their area. Securing this community participation is often dependent on the commitment to sharing the benefits of ecotourism development. The key issue here, then, is one of distributive justice, but there are obvious links here too with the kind of communicative ethics model proposed by Habermas which encourages those involved to talk through problems (see Chapter 5). Ideally, everyone (including the environment itself) benefits, but especially (as Rawls's theory demands) those in the most disadvantaged position, because the costs of setting aside local environments for conservation are offset through the direct provision of revenue and other non-material benefits to communities. These benefits should be an additional form of support for local economic activities, so that they complement rather than replace traditional local practices (Wallace and Pierce, 1996: 846–860). The handing over of conservation and ecotourism operations to sub-state entities, such as local communities, requires the development of dynamic and enthusiastic institutional arrangements. Since only a few ecotourism schemes have strong institutions that are capable of managing complex businesses for community development, it is not suited to every situation (Duffy, 2002b: 98–126; Steele, 1995: 34–36).

While community-based ecotourism implies a high degree of public participation, critics have pointed out that communities are very rarely given the chance to respond meaningfully to schemes that are supported by governments and/or the private sector. In this way, community-based ecotourism can often benefit only a narrow elite because the political nature of decision-making processes can often cut out communities and their interests. As a result, many community schemes have developed a tokenistic form of participation completely at odds with the Habermasian notion of an 'ideal speech situation' (Hall, 1994: 167–171). Such minimal participation then allows tour companies to package communities as a tourism attraction, with depictions of local people as smiling and welcoming faces for international visitors. Yet aspects of indigenous society and politics are kept away from the tourist gaze, so, for example, indigenous groups that assert land claims in tourist areas are redefined as disruptive and disloyal by central governments and the private tourism industry. This is because tourism development can become a struggle in which one powerful interest group attempts to legitimate its understanding of the appropriate use of space and time, while another, less powerful group resists this control (Hall, 1994: 182–200; Van den Berghe and Flores Ochoa, 2000). The example of community-based ecotourism in Belize will demonstrate how the broader politics of Belize and the relationships between indigenous Mayans and the central government have a critical bearing on its workings.

The politics of community conservation in Belize

As Chapter 6 highlighted, the Belizean government has been actively involved in promoting the tourism industry.

The development of community-based ecotourism has a specialist niche in the broader tourism industry, offering visitors alternative forms of accommodation such as basic guesthouses and homestays that are locally owned and managed in a way that spreads the income to the whole community (Moscardo *et al.*, 1996: 29–33). Southern Belize is particularly well known for community-based ecotourism ventures that are part of wider conservation schemes and reliant on the idea of local participation and management. There are a number of these initiatives in Belize that centre on conserving resources as diverse as bird life in Crooked Tree, howler monkeys in the Community Baboon Sanctuary, jaguars in the Cockscomb Basin Preserve, manatees at Gales Point, and Mayan ruins in the El Pilar Archaeological Reserve. These schemes are informed by the debates about community-based natural resource management (CBNRM – hereafter referred to as community conservation). Supporters of community-based ecotourism development suggest that protected areas with genuine community participation in ecotourism can form the cornerstone of development plans at the local, national and regional levels (Place, 1991: 196–201; Wilkinson, 1992: 386–392). Such tourism is purported to be 'ethical' not only in terms of distributive justice, but also in the sense that success is critically dependent on ensuring that the *rights* of indigenous peoples are respected.

The Tourism Strategy for Belize (more commonly referred to as the Blackstone Report) identified the development of community-based initiatives and micro-enterprises as part of the key to a successful ecotourism industry. The Blackstone Report recommended that the stakeholders involved in community-based eco-tourism should agree to the ecotourism initiatives in the first place and the benefits should be shared among all levels of society. This was intended to attract more ecotourists to Belize, to create a successful industry. In particular, Toledo District was highlighted as a possible eco-cultural zone to attract international visitors, where a Belize Maya Heritage Trail could be used as a marketing tool (Ministry of Tourism and the Environment/Inter-American Development Bank, 1998: 1–10). The geographical remoteness of parts of Toledo District, its relative lack of development, and its reliance on subsistence agriculture and small-scale revenue-generating ventures has led to a different kind of ecotourism. The identification of Toledo District, where the Mayan community constitutes the majority of the population, as a key part of the national tourism plan was also ethnically and politically significant. Toledo District is particularly known for its Mayan village homestays, where ecotourists are encouraged to spend time in a Mayan village as part of a cultural tour of indigenous peoples. The Toledo Ecotourism Association (TEA) was established partly in response to the experience of other communities with ecotourism to ensure that communities retained revenues and other benefits from ecotourism ventures. Most villagers have little or no experience of ecotourism and they lack the necessary capital and training to ensure that ecotourism in their area is a commercial success. However, with community-based ecotourism the objective is to ensure that local people retain control and exercise real decision-making power.

The TEA programme is a communally managed project that provides separate guesthouses within Mayan villages. The villages involved in the TEA share the revenue from all the activities surrounding ecotourism, such as guiding, accommodation, provision of meals and entertainment. For example, in TEA villages such as Medina Bank (Plate 7.2), a different family provides meals for visitors each day and is directly paid for inviting ecotourists into its home. In addition, members of the community have been sent on courses to enhance their hospitality skills and to discuss future options for developing ecotourism for the benefit of the community (interview with Rafael Cal, TEA chair, Medina Bank, 10 January 1999).

Participation in the TEA scheme clearly differentiates it from the other eco-tourism initiatives in the district. Each participating village has an elected representative (*alcalde*) in the TEA project to ensure that the interests of all the communities are represented in the schemes. The success of the TEA programme has won it international attention. In fact, the TEA guesthouse programme won the

Plate 7.2 Rafael Cal of the Toledo Ecotourism Association, and a TEA guesthouse in Medina Bank Village, Belize, 2000

world prize for socially responsible tourism in 1996, presented at the International Tourism Exchange trade show in Berlin (*Amandala*, 1997b; *San Pedro Sun*, 1997a; Schmidt, 1999).

Laguna was the first village to complete its guesthouse in 1991, followed shortly after by other TEA villages such as San Pedro Columbia, Barranco, San Miguel and Santa Cruz. Laguna village hosted its first set of ecotourists in 1991, and from then on visitors were sent to the villages on a rotating basis in order to spread the income. The trips were specifically marketed to visitors who had an interest in local cultures, did not require luxurious accommodation, and wanted to explore pristine natural environments (Williams, 1993b: 3–8). For example, some of the villages offered trips to see nearby waterfalls, unexcavated Mayan ruins, horse-riding trails, medicinal plant trails in the rainforests, and tours of herbal or botanical gardens in the villages. Furthermore, some of the participating villages had considered the development of eco-trails that incorporated two- and three-day hikes into the Maya Mountain Reserve with villagers acting as guides and porters (interviews with Rafael Cal, TEA chair, Medina Bank, 10 January 1999; and Pio Coc, Toledo Maya Cultural Council, Punta Gorda, 24 May 2000). Chet Schmidt, an adviser to the TEA, suggested that the majority of ecotourists who were interested in the Maya village stays were upper middle class, well educated, experienced in travelling in the developing world, and concerned to care for the environment. In addition, he indicated that such visitors were willing to pay more for the TEA scheme than for other, similar trips because theirs was more socially responsible and culturally aware than others (interview with Chet Schmidt, TEA and Nature's Way Guesthouse, Punta Gorda, 8 January 1999). One indicator of the types of traveller that the scheme appeals to is the fact that it was nominated for a sustainable tourism award by the up-market *Condé Nast Traveler* magazine. The TEA programme also received support from parts of the local hotel sector and assistance from other businesses because it would attract more tourists (and their spending power) to southern Belize, and especially Punta Gorda.

However, ecotourism development for Mayan communities in southern Belize has not been unproblematic. Like all tourism developments, it has benefited some sections of the local community more than others (see Van den Berghe, 1994; Van den Berghe and Flores Ochoa, 2000). Participating communities have experienced differential benefits according to their level of local organization, ability to lobby within the TEA, and even their geographical position. For example, Medina Bank is situated on the southern highway and has very good access to national bus networks, while San Pedro Columbia and Barranco have less frequent public transport. As a result, some ecotourists, without their own transport, are keener on villages that have good links, because remote villages raise problems for onward travel for visitors on a tight schedule. In addition, the question of genuine community participation in ecotourism is also problematic. For example, many of the villages in Toledo District do not have the technical capacity to develop and manage bank accounts and deal with complex financial transactions. Similarly, the remoteness of some villages from banking facilities is also a barrier to communities being able to exercise full financial control over the economic benefits of ecotourism (interview with Gregory Ch'oc, Kekchi Council of Belize, Punta Gorda, 23 May 2000).

While community-based tourism is often presented as a means of providing sustainable development for Mayan communities, some Mayan leaders have questioned the effectiveness of tourism initiatives. Gregory Ch'oc of the Kekchi Council for Belize pointed out that ecotourism was not the solution to the problems facing Mayans in Central America. The council is more concerned to develop a range of economic activities for Mayan communities, and ecotourism would be one activity among others. In particular, there has been an attempt to develop sustainable forestry (focusing on tropical hardwoods) and organic farming of maize and cacao, both areas where Mayan communities already have substantial experience, and which therefore require few resources to establish. Toledo District is already nationally known for rainforest products such as cashew fruits and nuts, medicinal herbs and organic cacao. This contrasts with a reliance on ecotourism or community-based tourism, which can take a number of years to show any kind of economic profit, or even social and environmental benefits.

This economic diversity can itself prove a barrier to full communal participation in ecotourism. In Toledo District, and especially for Mayan communities, tour guiding provides only part-time employment and is one of a number of economic activities that are undertaken by individuals. Guiding may be the income earner for one day per week, but the other six days will be devoted to subsistence agriculture, hunting and fishing. However, the Belize Tour Guide Association rules require all guides to attend relevant courses and to pay an annual fee to remain licensed (and legal) guides. This was part of a broader programme to pro-fessionalize the industry. For part-time guides in Toledo District this constituted a real barrier, because they did not earn the same levels of revenue from ecotourism as guides on the (more popular and more expensive) cayes, and they could not devote time to residential courses in Belize City if they were tied to subsistence agricultural systems (*Amandala*, 1999g; *San Pedro Sun*, 1997a).

In response to complaints about the cost in time and money, the Belize Tourist Board and the Ministry of Tourism agreed that TEA members could be called Village Site Guides. The licence to be a Village Site Guide cost only B$10 (US$5) and did not require attendance at formal training sessions in Belize City. However, this formal acknowledgement of the different circumstances for Mayan people led to complaints from members of the conventional tour-guiding sector, who had to pay B$140 (US$70) for their licences, attend courses in Belize City and prove they had no criminal convictions. They claimed that the TEA guides were being given unfair preferential treatment and that a number of them had convictions for growing marijuana (*Amandala*, 1997a, d). There is an ethical issue here in terms of whether the regulatory framework should espouse equality in treatment or try to enact some form of distributive justice by varying fees for those in materially different circumstances. In any case, it is clear that despite winning global awards, ecotourism initiatives in southern Belize were still facing problems and some opposition from other parts of the Belizean ecotourism industry.

The development of community-based tourism in Toledo District was further complicated by the demands for political recognition of land claims by indigenous peoples in the region. Community-based ecotourism has thus become intimately bound up with one of the most politicized issues in Central America, and any discussion of sustainable development needs to be tempered with a recognition of the political constraints within which it has to work. The Mayan communities in Toledo District have consistently lobbied for the establishment of a Maya homeland, which is related to calls for the development of a Maya eco-park. The proposals for the eco-park include provision for a controlled amount of ecotourism to ensure that it is self-financing. According to Pio Coc, head of the Toledo Maya Cultural Council, the idea of the Maya homeland began with a map drawn by Mayan communities showing the places that they already lived in and used for agriculture and hunting. At the moment, the land that Mayan people occupy in Toledo District is a government reservation. As a result, they feel they have no rights of ownership and that the land could be taken away from them at any point (interview with Pio Coc, 24 May 2000; Toledo Maya Cultural Council and Toledo Alcaldes Association, 1997).

The demands for a Maya homeland have been resisted by the central govern-ment because of the implications for further claims for land and political rights in Belize and the wider Central American region. The Toledo Maya Cultural Council was concerned that the Maya homeland demand was being frustrated because the central government in Belize believed that Mayans were asking for a state within a state. Chet Schmidt, of the TEA, claimed that the central government was blocking the eco-park because of anti-Mayan racism and because the Mayans in the area had a history of anti-government activities. In addition, he suggested that a success in the region would draw attention to the political corruption that had swallowed up previous funds destined for the area. He also argued that there was an active attempt to prevent ecotourism in the Toledo District because the north and west of the country already had well-developed tourism infrastructure. The powerful interest groups in the established ecotourism areas (including tour operators, hotels and bars) wanted to retain control over the flow of visitors in the country (interview with Chet Schmidt, TEA and Nature's Way Guesthouse, Punta Gorda, 8 January 1999).

It is clear from an examination of the ecotourism initiatives in southern Belize that the ethics of sustainable development is highly complex. On a conceptual level, the lack of any clear definition of what sustainability means has led to its being used to label a broad range of environmental schemes and initiatives. The ecotourism programmes in Toledo District are often justified and promoted through reference to the need for sustainable development. However, it is clear that the reliance on ecotourism intersects with the dominant discourses of other interest groups, such as the neoliberal international financial institutions, environmental NGOs, government agencies, and so on. The ethical issues that emerge can be understood, and are sometimes voiced, in terms of a variety of competing ethical discourses, including claims to indigenous land rights by Maya campaigners and arguments about whether or not tourism is actually the best way to maximize a community's utility. Even where debate could be carried out within a single discursive framework, such as utilitarianism, there is still plenty of scope for differences to emerge. Perhaps forestry, organic farming or a mixed economy might prove more beneficial, at least in the short term, to many local communities, while some have argued that sustainable tourism offers a long-term future. Such issues lead us back into debates about the difficulties of making utilitarian calculations and predicting consequences (see Chapter 3). The issue of distributive justice among different ethnic groups and economic and political alliances is also obviously extremely important and potentially explosive. Even the apparently small matter of guide fees threatens to upset sections of the ecotourism industry when people feel they are being treated unfairly. On yet another level, the Toledo project itself might be seen as an attempt to put into practice the notion of an ideal speech situation (Chapter 5) where the voices of all concerned might be heard. But as we have seen, voices and power relations are never equal, and the power of nationalist politics and global capital often threatens to drown out those most intimately concerned with sustainable development initiatives, namely the Mayan communities themselves. So long as ethical values continue to play second fiddle to economic values on the national and global stage, then even the best initiatives remain at risk from developments that exploit local people and environments for short-term profits.

CAMPFIRE: ethics and animals in community-based tourism

We have looked at some of the ethical issues that have arisen in discussions about how the natural environment might be used for community benefit and tourism. However, while debates about community-based tourism projects might seem to be mainly about human rights, utility or distributive justice, they are further complicated by ethical questions concerning our 'right' to treat animals and the environment simply as resources for the benefit of human communities. One of the problems with so many of the ethical frameworks we have examined has been the fact that they ignore or exclude our ethical relations to our wider non-human environment. While 'nature' does have its advocates, especially in the form of various conservation lobbies, it appears in sustainable development's discourses largely as a resource, as something to be made use of simply for human ends or purposes. This is of course (as Chapter 1 argued) the very antithesis of an ethical approach, which, at least in non-utilitarian forms, should treat things as 'ends-in-themselves' (Chapter 4) or as Others deserving respect (Chapter 5). Such issues in environmental ethics (Hargrove, 1989; Nash, 1989; Smith, 2001a) and what is loosely called animal 'rights' (loosely because some of the most prominent advocates for the ethical treatment of animals, such as Peter Singer (1991), are actually utilitarians) add yet another layer of complexity. This section will examine the complex set of ethical issues that arise from a programme that attempts to satisfy the needs of rural people through the definition of wildlife as a resource to be used for human benefit.

During the 1990s Zimbabwe's Communal Areas Management Plan for Indigenous Resources (CAMPFIRE) became internationally renowned for its efforts to reconcile the needs of conservation and development. It was promoted as a model programme for community-based natural resource management that included local control over tourism developments. Begun in 1986, it aimed to ensure that the rural communities living in Zimbabwe's semi-arid and marginal Communal Areas were able to capture the benefits from wildlife utilization. In this way CAMPFIRE was intended to strike a workable and ethical balance between wildlife conservation and meeting the basic needs of rural people. However, CAMPFIRE is informed by a complex set of ethical norms that are sometimes in direct conflict with values generally associated with environmental ethics or animal rights. This, then, leads on to a tension in any discussion about *ethical* tourism.

CAMPFIRE tourism heavily relies on sport hunting, especially of elephants. This is defined as consumptive tourism, which means that small numbers of certain species can and should be killed in order to pay for conservation of a species as a whole and for rural development initiatives such as the building of schools and grinding mills, and the drilling of wells. Zimbabwe's wildlife policy, which is intimately linked with the tourism and rural development sectors, is based around the idea of sustainable utilization, a policy that defines wildlife as a commodity that can generate revenue through a series of activities. These activities range from photographic tourism, sport hunting, sales of meat and hides, to auctioning wildlife to the private ranching industry. In the context of ethical tourism, this obviously raises an additional set of debates centred on the relative ethical value of animals and humans.

Tourism, Zimbabwe and CAMPFIRE

In Zimbabwe, conservation and tourism policy is informed by a commitment to sustainable utilization of wildlife to meet the basic needs of people, especially rural communities. It is also intended to satisfy ethical concerns around balancing out human development, conservation and the humane treatment of animals. This sustainable utilization approach can be divided into two categories: consumptive and non-consumptive uses of wildlife. Non-consumptive uses of wildlife include activities such as photographic tourism and birdwatching, while consumptive uses of wildlife involve the use of animals and their products through the sale of meats and hides, as well as sport hunting for trophies such as horns and ivory (see Plate 7.3). CAMPFIRE governs wildlife policy in Communal Lands where wildlife densities are too low to support the photographic safaris that are more commonly associated with other African destinations such as Kenya and Tanzania (Sindiga, 1999), and this means that CAMPFIRE areas are especially reliant on commercial elephant hunting.

The Communal Lands (or Communal Areas) of Zimbabwe are rural areas that were originally established by the white settler regime in the pre-independence period. They were areas allocated to indigenous Zimbabweans, while the more

Plate 7.3 Elephants at Hwange National Park, Zimbabwe, 1996

fertile agricultural lands of the highveld were allocated to white commercial farmers, who concentrated on tobacco and beef production. They are characterized by poor agricultural land, subsistence farming and higher population densities than the commercial farming areas. In general, they were the least developed areas of Zimbabwe, and at independence the new Zimbabwe African National Union (ZANU) government pledged to redress the colonial imbalances through rural development initiatives and a land redistribution scheme (Drinkwater, 1991; Stoneman and Cliffe, 1989). As one of these development programmes, CAMP-FIRE must be seen in the light of Zimbabwe's post-independence politics and the broader global political and economic scene.

Africa is estimated to have received around US$10 billion worth of the US$476 billion global tourism industry (WTO, 2001). Much of this is associated with wildlife tourism, and Zimbabwe, like other nations, relies on visitors coming specifically to view animals in their natural habitat or see landscapes like the Victoria Falls. Many seek the *authentic* safari experience (see Chapter 6). The Zimbabwe Tourism Development Corporation (ZTDC) noted that tourists were motivated to visit by notions of exotic appeal, uniqueness or a sense of 'real Africa', the friendliness of the people, good infrastructure and security, a cultural or historical dimension to the trip, and value for money, with the most popular trips being game safaris (Dann, 1996a; ZTDC, 1993). Developing states containing natural areas with little conventional tourism development have turned to eco-tourism as a means to earn foreign exchange while ensuring that the environment is not degraded (Boo, 1990: 1–3). This revenue-generating capacity mitigates some of the very high costs of protection and maintenance of national parks and wildlife in developing states (Zimbabwe Trust, 1992).

Until the recent political crisis, the tourist industry was a major revenue earner for Zimbabwe, and in 1993 tourism was the third foreign exchange earner after agriculture and mining, but with droughts affecting agriculture in 1994 and 1995, tourism became the second largest earner. In 1995 tourism reached a new peak, with a record 1 million visitors (*Financial Gazette*, 1995b; WTO, 2001). But after sustained growth, Zimbabwe as one of Africa's biggest destinations began to stagnate, with visitor arrivals declining from 2,103,000 in 1999 to 1,868,000 in 2001 as a result of expanding political violence and more general economic decline (*The Times*, 2001). Widespread reporting of political violence in the run-up to the 2002 elections exacerbated this tendency. This has had an inevitable negative impact on tourism initiatives in CAMPFIRE areas, which had always been a small and specialist sector of the wider tourism industry anyway.

Until recently Zimbabwe has been well placed to take advantage of the growing ecotourism market, with small-scale wildlife safaris on privately owned ranches and on rural community lands. The Parks Department *Research Plan* makes it clear that tourism is perceived as the key to the economic survival of the Parks and Wildlife Estate and to justifying wildlife utilization as a form of land use (DNPWLM, 1992a: 43). Attempts have been made to place economic values on the viewing of different species, and such valuation has been most developed in Kenya, where it was estimated that by 1989 the viewing value of elephants was already approximately US$20–30 million and increasing (Leader-Williams, 1994: 63). This was because the Western tourists who make up the bulk of Kenya's tourist market placed seeing an elephant high on their list of priorities in a wildlife safari. In Zimbabwe the consumptive values of wildlife have also been developed through sport hunting, and this means the economic values of wildlife are enhanced because they can be 'sold' twice: first for photographic safaris and then for sport hunting (Wildlife Society of Zimbabwe, 1993). The question remains, however, as Chapter 1 argued, whether this purely economic evaluation is capable of expressing the 'real' value of the wildlife or the values of tourists and local communities.

Since wildlife is, in any event, critical to the economic profitability of the tourism industry, many operators consider long-term planning for conservation as essential. For example, one of the major hotel chains, the Zimbabwe Sun Group, recognized that 'the long term success of this industry is totally dependent upon the harmonious interface between tourism development and the environment in which that development takes place' (Smith, 1994: 23). The Zimbabwe Sun Group argues that the environment will be conserved only if it is seen as a significant economic resource. It claims that its own development projects in Katete at Kariba Dam and Mahenye on the borders of Gonarezhou National Park were intended to provide major and tangible economic benefits to the company and to the people in the locality (ibid.: 24–25). Tour operators do not simply argue in favour of more effective conservation; they assist in very direct ways by providing an important source of funding and capital goods for some areas of wildlife conservation. Safari and tour operators often make significant contributions to the areas in which they operate, particularly in the form of road maintenance and providing diesel for vehicles (Cunliffe, 1994). For example, a private-sector organization, the Zimbabwe Council for Tourism, donated Z$2 million to the Parks Department in 1993 because it was concerned at the lack of funds in the department. The tourist industry acts as a vitally important support base for state-sector involvement in wildlife conservation. The industry support stems directly from its recognition that continued survival depends on the conservation of wildlife in the national parks and safari areas. Of course, it also provides private tour operators with good publicity.

There are, however, a number of problems associated with tourist development. Zimbabwe's policy of high cost–low impact tourism is intended to avoid some of the difficulties associated with mass tourism as found in East Africa, such as overcrowding of visitors in national parks, unplanned and uncontrolled building of tourist facilities and resorts, and the adverse publicity that tourist developments attract. The problem of visitor carrying capacity is faced by all states attempting to develop their tourist industry (see Sindiga, 1999). In Zimbabwe that capacity is relatively low, and so advertising and pricing mechanisms concentrate on attracting a small number of visitors, because the industry and the facilities in national parks and other scenic areas would be unable to cope with the demands of mass tourism.

However, the presence of even a small number of tourists raises potential ethical, cultural and class problems, such as the question of whether separate facilities should be provided for overseas visitors in national parks (Cumming, 1990a). The Parks Department's policy is to have a three-tier system of charges for entry into national parks: one price for overseas visitors, one for regional tourists (South Africans) and one for Zimbabwean citizens. The department argues that Zimbabweans contribute to the maintenance of national parks through taxation (for example, in 1995 Zimbabweans contributed Z$44.8 million in tax revenue to the Parks Department) and so are entitled to a discount (*Daily Gazette*, 1993; *Herald*, 1993, 1995). This in effect means that most local people are given a real and usable right of access to their own natural heritage. Of course, any level of entrance fee will prevent some members of a society from gaining access, but the multilayered fee system is meant to take account of global inequalities in income distribution. The underlying sentiment of this policy is that overseas visitors who have been able to afford a safari holiday can clearly afford to pay a small entrance fee as well. In accordance with this, the private-sector tour operators and hotel operators offer a range of facilities that are open to locals and overseas visitors. These policies ensure that Zimbabwean tourism attempts to answer the criticism that tourism excludes local people in order to privilege wealthy overseas tourists (Ap, 1992; Munt, 1994a, b).

There are, however, serious problems with reliance on tourism to fund development and conservation. Tourism can be fickle, since a steady flow of tourists relies heavily on image (Dann, 1996a). National parks, in particular, continue to rely on romantic imagery of wilderness and wildlife to attract visitors, but this imagery could easily be spoiled by large crowds, so protected areas have to restrict access by either cost, relative inaccessibility or quotas. On the other hand, globalization of Western culture is perpetuated through the tourist industry, which carries with it implicit representations of Africa as a gigantic outdoor zoo (Zimbabwe Trust, DNPWLM and the CAMPFIRE Association, 1990). Reliance on tourism to pay for conservation is also problematic because areas of great conservation importance, such as swamps, are not necessarily major tourist attractions (WTO *et al.*, 1992: 6–17).

Moreover, as we have seen, the growing civil unrest in Zimbabwe from the late 1990s has had a negative impact on the tourism industry. Political violence has created adverse publicity that easily creates anxiety among potential holiday-makers. Since Zimbabwe offers a wildlife 'product' that can be found in other southern and eastern African states, potential customers for photographic wildlife safaris can switch to Botswana, Tanzania or Kenya. Likewise, sport hunters can substitute Namibia, Botswana or South Africa for Zimbabwe. If, in a climate of political violence, tourism ceases to be a major foreign exchange earner, then the industry's underpinning of state and community-oriented conservation efforts will be in jeopardy (Bloom, 1996; Hall and O'Sullivan, 1996; Norton, 1994; Sonmez, 1998; Sonmez and Graefe, 1998).

CAMPFIRE, 'resource' utilization and development ethics

Its supporters often present CAMPFIRE as a new form of ethical tourism that ensures that local communities are the main beneficiaries of any new developments. However, since CAMPFIRE is highly dependent on the sport hunting of elephants and other consumptive uses of wildlife to generate revenue for rural development schemes, its critics argue that it fails to fulfil a commitment to humane treatment of animals. Supporters and critics of CAMPFIRE have presented the satisfaction of basic human needs and the intrinsic value of animals as two opposing and inherently incompatible concerns. This is not necessarily the case, since many environmental ethicists are more concerned with the intrinsic value of entire habitats and species than with individual animals. There is also a pro-hunting lobby within environmental ethics that regards hunting in certain circumstances as a 'natural', 'authentic' and necessary aspect of human life (Leopold, 1949; Shephard, 1997). The fact remains, though, that neither of these groups would be likely to support trophy hunting or the commodification of animals involved in CAMPFIRE's schemes. There is also something rather distasteful about relying on the colonial tradition of the 'white' hunter and the upper-class shooting party to fund supposedly egalitarian and anti-colonial development projects. Ironically, given that it is mainly white tourists who come to Zimbabwe to shoot wildlife, it could be argued that CAMPFIRE is merely a continuation of an imperialist, Western and anthropocentric agenda of game preservation, albeit with a wider distribution of economic benefits.

However, we should also note that the very existence of national parks in Africa and Asia and (to a lesser extent) the existence of the conservation movement in general are historically interconnected with an imperialist agenda of 'game preservation'. By the end of the nineteenth century the destruction of wildlife through over-hunting and diseases like rinderpest that followed in colonialism's wake meant that big game animals were scarce. Hunting preserves were set up by colonial administrations based on the transfer of property rights over game to white settlers, 'the exclusion of Africans from hunting, and the progressive conversion of the game from a direct economic resource in ivory, meat, hides and skins into an indirect one, a means of raising revenue from "sport" and tourism' (Brockington, 2002; Leach and Mearns, 1996; MacKenzie, 1988: 201).

Supporters of CAMPFIRE point to the ways that it has substantially increased the area allocated to wildlife. This has been central to the arguments to gain political and economic acceptance for a workable wildlife policy in the context of local demands for land redistribution. While the Parks and Wildlife Estate covers just over 12 per cent of Zimbabwe's land area, when CAMPFIRE areas and ranches under wildlife are added, the area of land under wildlife reached 33 per cent, at least until the recent land invasions by so-called war veterans, continuously growing (interview with Brian Child, Head of the CAMPFIRE Co-ordination Unit, DNPWLM, Harare, 16 May 1995; see also Compagnon, 2000; McGregor, 2002; Raftopoulos, 2002). Unsurprisingly, the first Communal lands to choose CAMPFIRE were those already relatively rich in wildlife. However, CAMPFIRE areas have bought wildlife to restock their lands in order to benefit from increased hunting quotas or tourism opportunities (*Herald*, 1989; *Sunday Gazette*, 1995).

Murphree (1995b) suggests that in Zimbabwe CAMPFIRE adopted an approach described as adaptive management, meaning that what worked was continued, what did not work was discontinued. The experience of commercial farms informed the decision to confer strong property rights on rural communities, but it was recognized that land units in the Communal Areas were not analogous to those in commercial farming areas. The then Director of the Parks Department pointed out in 1995 that in Communal Areas the land is not owned by anyone, but is utilized by the people on it. In the same way, wildlife is not owned by the community but is utilized by it, and so wildlife is a resource held in a group system that is governed by usufruct rights (interview with Willie Nduku, Director, DNPWLM, Harare, 7 July 1995). Problems with defining ownership of wildlife meant that it was difficult to decide who or what constituted an appropriate authority over wildlife. This was why the District Council rather than the individual farmer was established as the focal point of the initial institutional structure of CAMPFIRE in Communal Lands.

Common resource management is problematic where the resource and its appropriators are poorly bounded (Metcalfe, 1992a: 36). Zimbabwe is fortunate in having clearly defined ward and district boundaries around communities in rural areas, and built on this by introducing a new system of group ownership with strictly defined rights of access to wildlife for the communities resident in that area (Murombedzi, 1992: 13). The Ministry of Environment and Tourism (1992: 4) claims that this represents 'a far sighted approach which recognises that landholders should be the best custodians of their natural resources provided they have the right to use wildlife and to benefit from their custodianship'.

Sport hunting as ethical tourism?

Sport hunting tourism in CAMPFIRE areas was based on the experience of private wildlife ranching in Zimbabwe. Wildlife ranching has its origins in the 1960s when legislative changes encouraged commercial farmers to conserve wildlife, but it was the 1975 Parks and Wildlife Act that proved to be the major catalyst in private-sector wildlife conservation. Under the Act, private landholders received appropriate authority status, which gave them the right to use wildlife (Child, 1988: 182–336). In 1994 it was estimated that 75 per cent of private ranches derived some or all of their income from wildlife-based ventures and that the area allocated to wildlife on ranches was expanding at 6 per cent per annum during the 1990s (Martin, 1994a: 7). In the 1990s the value of wildlife on ranches was greater than that of cattle and crops, especially in ecologically marginal areas such as the lowveld in the south-east (Jansen, undated; Zimbabwe Trust, 1992: 31–46).

There has been a growing demand for tourist facilities that do minimal damage to the environment, and African wildlife holidays have proved particularly attractive. In the short term, sport hunting by overseas clients provides the greatest returns with minimal investment in facilities for visitors. This is because hunters generally require only basic infrastructure, whereas tourism needs a level of investment to establish international-standard accommodation and roads to carry viewing vehicles. Non-consumptive photographic tourism has been developed in some CAMPFIRE areas, but this is feasible only where there is a significant scenic or sporting attraction such as white-water rafting in the Zambezi River and Victoria Falls area. As well as this, rural communities have established some smaller-scale tourist facilities largely aimed at the cultural tourism and ecotourism markets. For example, Sunungukai tourist camp in Hurungwe District offers safaris to local and foreign tourists. The community also provides basic accommodation in chalets and offers guided walks, fishing, visits to local villages, and local handicrafts. Since there is no commercial operator involved, the community runs the chalets and retains all the revenue they generate (interview with Keith Madders, Director, Zimbabwe Trust (UK), Epsom, 12 October 1994). This kind of non-consumptive development is, however, the exception rather than the rule because communities see an immediate return from the trophy fee paid by the hunter. A single foreign hunter can bring in up to seventeen times the foreign currency that an individual photo-safari client brings in. This in turn means that communities can see a direct link between wildlife and revenue (Jansen, undated). Supporters of sport hunting also argue that a hunting party which kills a single elephant will create much less stress on the environment than numerous groups of tourists in vehicles interested in wildlife viewing.

The fact that private and state interests are involved in revenue generation, institution building, funding and representation of rural interests in CAMPFIRE means that the programme enjoys considerable layers of support, which assists the Zimbabwe Parks Department's political justification of its policy of sustainable utilization. The department states that hunting has been limited to small numbers of suitable species in order to ensure that wildlife is used sustainably. It sets a hunting quota for each CAMPFIRE area, which is based on surveys of wildlife populations, though hunting quotas can be increased slightly by allowing 'problem animal control' to be carried out by commercial safari operators. Hunting rights are sold to safari operators who pay for the right to hunt and the fees for each animal shot on the quota. For example, Tshabezi Safaris paid Muzarabani District Council the equivalent of US$108,025 in one year in hunting and trophy fees (Child, 1995d: 1). Yet rural communities do not choose the safari operator solely on the level of fees offered. In some areas the community will choose a tender that also offers incentives, such as the building of a road or lodges. In some Communal Areas tenders were awarded on the basis of an existing good record of co-operation between an established safari operator and the community (interview with Sally Bown, Zimbabwe Association of Tour and Safari Operators (ZATSO), Harare, 9 March 1995). In order to ensure greater local control over hunting firms, guidelines were drawn up by government agencies and NGOs involved in CAMPFIRE on how to run joint ventures between district councils and private operators (Jansen, undated: 4).

Safari operators' central role in revenue generation for CAMPFIRE areas means that they exercise considerable control over how wildlife resources are utilized.

This has not been unproblematic: some communities attempted to set up their own safari operations so that the community could retain all the fees and profits. But overall these have not been successful because the councils lacked the necessary expertise. For example, the community-run safari operation in Guruve failed to compete with the returns offered by commercial operators. Keith Madders suggested that the failings were due to lack of experience and the expectations or perceptions of potential clients. Part of the hunting experience is the nightlife – that is, swapping stories around the fire, where overseas clients appreciated someone well travelled (and white). Rural people could not compete with the services offered by commercial operators, such as catering for the dietary preferences of overseas clients, professional hunting equipment, the 'right kind' of social life, and tented accommodation (interview, 12 October 1994). The fact that the safari industry is white dominated has led to criticisms of CAMPFIRE. Critics suggested that CAMPFIRE has protected white safari operator interests over those of rural people, and that safari operators made very large profits from CAMPFIRE, which explains their vocal support for the policy (Murombedzi, 1992). This kind of distributive issue mirrors tourism development in rural communities in general. Critics have been concerned that the largest profits have gone to international tour operators and hotel chains rather than rural communities.

The starting point for Zimbabwe's conservation policy is that natural resources exist in a social, economic and political context. CAMPFIRE constitutes an area of policy where there is a clear focus on the political nature of conservation, in the face of claims that conservation is about the science of saving animals. Zimbabwean state agencies and NGOs have sought to use the rhetoric of sustainable utilization to bring political standpoints and moral questions to centre stage in the conservation debate. The idea of sustainable development and, in particular, the sustainable use of natural resources is intended to remove the conflict between environmental care and economic development. This approach draws on the concept of a safe minimum standard of utilization of resources in order to ensure that the natural resource base is not depleted faster than it is replaced. Natural resources are defined as all forms of matter and energy available to the nation for development and enjoyment of people. Conservation is similarly defined as the process whereby natural resources are managed in such a way as to ensure sustainable use and development, a process whereby the environment is modified to meet human needs and enhance quality of life (Ministry of Natural Resources, 1990: 4).

To some extent, sustainable utilization is a pragmatic solution. Metcalfe (1992a: 10) argues that since governments in developing countries faced difficulties in managing protected areas without the co-operation of local communities, wildlife policy should address questions of economic efficiency, environmental integrity, and equity and social justice. Sustainability requires an expressly political solution as well as the technical prescriptions of environmental science, and it is recognized that wildlife cannot be conserved in the face of antagonistic social and economic forces. Hence the Zimbabwe Parks Department is concerned that wildlife conservation is likely to be successful only if wildlife can be used profitably and the primary benefits accrue to people with wildlife on their land (Ministry of Environment and Tourism, 1992: 2). Wildlife utilization is deemed to be the most effective means of establishing social and economic forces favourable towards conservation (Ministry of Natural Resources, 1990: 9). Zimbabwe's wildlife tourism policy clearly has to accommodate a number of competing interests, including protected areas, rural poverty, land pressures and lack of finance (Metcalfe, 1992a: 4; Zimbabwe Trust, 1992: 1). CAMPFIRE attempts to create a political settlement between potentially competing interest groups through the articulation of sustainable utilization as a political ideology and as a means of defining a practical wildlife management policy.

However, this settlement raises other issues, since it clearly reveals that the environment is conceptualized in anthropocentric terms. In the context of Zimbabwe, people remain a clear priority over animal welfare and conservation. This strong commitment to a particular economic and moral standpoint has meant that Zimbabwe's wildlife tourism policy has faced some criticism. The role of hunting operators has revealed clear divisions in the conservation community and has highlighted the fact that conservation itself always entails the articulation of particular sets of ethical norms. CAMPFIRE's reliance on hunting for revenue brings conflicting values about sustainable utilization into sharp relief from two angles. First, there are many conservationists who believe that treating our natural environment as a resource is part of the problem rather than a solution to our ecological crisis. Second, there are issues concerning animal welfare that focus on the actual practice of hunting.

The first group point out that there are serious ethical issues in treating the environment as nothing more than a human resource. They call on many different philosophical traditions, but all agree that the key problems of modern society are related to its unethical treatment and commodification of nature. They regard the fact that we are always willing to exploit the natural environment for our own instrumental purposes as a form of anthropocentrism (human-centredness) that fails to recognize that it has an intrinsic or inherent value of its own, whether or not we find it useful. They argue that we can have ethical, rather than purely economic, relations with elephants, lions, rainforests and perhaps even waterfalls and mountains (Devall and Sessions, 1985; Plumwood, 2002; Smith, 2001a). Some have termed this approach deep (as opposed to shallow) ecology, a term first employed by the Norwegian philosopher Arne Naess (1972). Deep ecology is a radical rather than a reformist ideology which 'offers a normative critique of human activity and institutions, and seeks a fundamental change in the dominant worldview and structure of modernity' (Katz *et al.*, 2000: ix). Such attitudes have considerable influence even on mainstream conservation biology (Kellert and Wilson, 1993; Soulé and Lease, 1995).

The second group deploys a variety of ethical arguments that focus on animal welfare. Some are couched in terms of animal rights (see Chapter 4), where at least some animals are regarded as possessing the requisite features to make us recognize our moral duty towards them (Regan, 1988). Others follow Bentham's utilitarian approach, which famously argues that those animals that can feel pain or pleasure must be included within any utilitarian calculus: 'the question is not, can they reason? nor can they talk? but, can they suffer?' (Bentham, 1907: 311). Thus Peter Singer (1986: 24–32) suggests that as intelligence and ability to communicate (verbally and non-verbally) does not entitle one human to exploit another, the same principle should be extended to animals. To fail to take account of the suffering of animals is, Singer argues (1986, 1991), equivalent to a form of speciesism – that is, irrational and unjustifiable discrimination against other animals by humans. These arguments do not necessarily mean that it is always wrong to kill animals (although Singer is a vegetarian), but they try to expand the boundaries of moral considerability beyond humans (for a critical appraisal, see Smith, 2001a: 30–44).

Such ideas are readily translated to animals like elephants, which are the object of CAMPFIRE's hunts. Animal liberationists point to elephants as an example of an animal that is capable of complex forms of communication, grieves the loss of family members, and creates close bonds with other elephants in its group (Jordan in Care for the Wild, 1992). Until now, CAMPFIRE has, rather curiously, been largely ignored by animal rights groups even though it is heavily dependent on sport hunting, but the industry still remains vulnerable to pressure from global campaigns. In 1989 approximately 63 per cent of all overseas sport hunters came from the United States, and this, together with other Northern countries, is where animal welfare groups are concentrated (interviews with David Cumming, Director, WWF-Zimbabwe, Harare, 22 February 1995; and Keith Madders, Director, Zimbabwe Trust (UK), Epsom, 12 October 1994). Reliance on sport hunting means dependence on overseas markets, which have been threatened by various moves inspired by the animal welfare lobby to disallow trophies from entering the United States. If Americans were not allowed to bring trophies home they might not be willing to pay the high fees that Communal Areas now rely on. The abuse of hunting permits is also a problem, since there is always a possibility that a hunter will shoot one trophy animal and shortly after see another with a better trophy. In some cases corrupt safari operators might be tempted to allow the hunter to shoot the second trophy and sell the original trophy for personal profit (Bonner, 1993; also see Duffy, 2000b: 38–41). The promotion of sport hunting also has the potential to discourage Western tourists who are interested in photo-tourism, since tourists can be encouraged to boycott states that do not adhere to strict wildlife preservation.

So, while conservationists are mainly interested in preserving elephants and other wildlife as a species or part of an ecosystem, the animal welfare lobby, especially animal rights activists, are concerned over the very idea of trophy hunting. From this 'rights' perspective CAMPFIRE relies on cruel practices that cannot be supported even if they contribute to poverty alleviation and rural development. We should, however, note that the arguments for preserving individual animals can sometimes conflict with conservation objectives. The idea of animal rights has been criticized on the grounds that such rights may conflict with other ecological principles that, for example, want to ensure maximum biodiversity. Utilization can be a means of ensuring that certain wildlife populations – such as elephant populations – that reach excessive numbers do not damage the environment and thereby reduce biodiversity. If elephants are allowed to multiply, other important species, such as rhino, may be forced out of an ecosystem owing to lack of food and water. Jon Hutton, Director of Projects, Africa Resources Trust (ART) (now ResourceAfrica) thus points out that it is important not to conflate individual animal welfare issues with conservation (interviewed in Harare, 13 March 1995).

The Zimbabwe Parks Department also points to the favourable status of wildlife populations as the evidence that Zimbabwe's conservation policy has been successful. Zimbabwe's problem with elephants is not that the country has too few, but rather too many. In the case of elephants it can be argued that since elephants are so numerous in Zimbabwe, their preservation value is very low (Barbier *et al.*, 1990: 21). In contrast, because there are so few rhinos, preservation value of the remaining ones is very high. However, whatever their differences, both animal welfarists and those conservationists who believe in the non-instrumental value of the environment agree that policies and philosophies that value animals in terms of economics and human use will lead to conservation of selected useful species to the detriment of others (Eckersley, 1992: 33–49). The Zimbabwean government admits that utilization of wildlife will affect population levels and composition, but does not accept that this is an argument against use as such. It argues that since wildlife is defined as a renewable (rather than a finite) natural resource, it can be used without destroying the resource base in the long term. Utilization of wildlife relies on an offtake which is proportional to the size of a population, and economic returns that are reinvested into the species have to be beneficial for its status (Martin, 1994b: 3–8).

It is elephants that have attracted the largest trophy fees. Sally Bown, of the Zimbabwe Association of Tour and Safari Operators (ZATSO), stated that in 1993 the fee for shooting an elephant in a safari area operated by the Parks Department was set at US$3000 whereas in a CAMPFIRE area anything up to US$9000 could be paid (in an interview, Harare, 9 March 1995). In Zimbabwe the sport hunting industry grew from a value of US$195,000 in 1984 to US$13 million in 1993, and the number of hunts rose from 25 to 1,300 over the same period (*Financial Gazette*, 1995a). This growth was assisted by the special reservations on sport-hunted ivory and other wildlife trophies that Zimbabwe lodged with the Convention on the International Trade in Endangered Species (CITES), as a result of which sport hunters can export their trophies from Zimbabwe to their home country if they have the relevant documentation (Craig and Gibson, 1993).

The issue of ivory has become a key area of conflict between the majority of Western conservationists who support an international ban on the ivory trade and those who see it as a valuable and potentially sustainable resource. Supporters of CAMPFIRE have criticized the international ivory ban as detrimental to community-based conservation since it reduced the yield per hectare of wildlife utilization schemes (*Herald*, 1994; *Sunday Mail*, 1990). Almost half of the 35 tonnes of ivory in Zimbabwe's central store came from Communal Areas, representing an opportunity cost to communities of US$3.5 to US$7.5 million if ivory is worth US$200–US$400 per kilo (prices based on offers made to the Parks

Department from various sources in the post-ivory ban period) (Child, 1995a). Consequently, agencies involved in CAMPFIRE have devoted time and effort towards political lobbying aimed at keeping existing trade links open (such as markets for sport-hunted ivory) as well as trying to reopen ivory trading. It is, however, important to note that CAMPFIRE has been successful in generating large amounts of revenue and has never been dependent on the ivory trade, despite claims by some Zimbabwean conservationists in 1989 that CAMPFIRE would fail if the ivory ban was implemented. Many conservationists also argue that lifting the ban would reopen a global trade in ivory, leading to an increase in (unsustainable) poaching in other countries where wildlife is less well protected. This might have catastrophic results on elephant populations outside Zimbabwe.

This conflict between ecological, economic and ethical values is reminiscent of the examples we examined in Chapter 1. Some environmentalists would argue that wilderness areas are priceless and that with utilization Zimbabwe is wrongly attempting to place a price on it (Sheldrick cited in Care for the Wild, 1992). Critics suggest that Zimbabwe's policy is an example of the commodification of nature, which means that wildlife and other environmental resources are valued only in terms of their economic worth. In effect, critics argue that wildlife has value beyond its economic worth since environmental, aesthetic and cultural values are critically important. Such ideas are sometimes extended to include the belief that human interference in the form of management of the environment upsets the 'natural' balance. It is suggested that humans are not given steward-ship over nature and do not have a right to interfere with it (Zimbabwe Trust, 1992), and that nature can correct its own imbalances and should be left to do so. However, preservation of wildlife within reserves might itself be seen as managing and interfering with nature, since humans too are a part of the environment.

The Parks Department argues that the zero-use option as represented by the preservation has failed across the continent. The reason, according to proponents of sustainable use, is that preservation was doomed to fail because it was based on an alien aesthetic framework and placed neocolonial moral judgements about wildlife into a legislative framework (Zimbabwe Trust, 1992: 1). CAMPFIRE is deemed to work by the Parks Department and associated NGOs because it harnesses and builds upon pre-existing African environmental values and tra-ditions rather than ignoring them (Kasere, 1995). We should note, though, that Zimbabwe has readily adopted a tourist industry that also relies on 'alien' economic and aesthetic values, so the arguments of anti-hunting conservationists are not alone in having colonial associations. It is true, however, as Marshall Murphree (1995b) argues, that it is hard for relatively well-off people in the industrialized world to understand the cost paid by rural people to maintain an ideal wilderness. Wildlife policy makers also argue that a hands-off approach may be disastrous when applied to ecosystems within African states where wildlife conservation has to compete with other forms of land use (Martin, 1994b: 11). Supporters of CAMPFIRE argue that Western preservationism is ultimately self-defeating since it negates the role of local people in wildlife conservation. If sub-Saharan African governments attempt to enforce a policy of zero use, utilization inevitably takes place on an illegal and unsustainable basis (Martin, 1994a: 8).

Lastly, we should note that CAMPFIRE exists in areas where land that is put under wildlife would otherwise imply forgone development opportunities for local people. Since wildlife entails so many social and economic costs for rural people, their approval for conservation policies is critical, and without it, conservationists in Zimbabwe argue, wildlife is doomed (Metcalfe, 1992: 1–7). Rowan Martin (Assistant Director of Research, DNPWLM, interviewed in Harare on 29 May 1995) argues that the 'use' of wildlife is unstoppable and the real ethical question is not whether wildlife should or should not be utilized, but whether it can be turned into a useful resource. Zimbabwe has therefore made a number of policy choices that attempt to reconcile the demands of conservation and international responsibility for its wildlife heritage, with possibly conflicting demands for economic growth and social and political acceptability (Zimbabwe Trust, 1992: 2).

CAMPFIRE represents an innovative approach to the difficulties of balancing wildlife conservation and rural development in Zimbabwe. However, its heavy reliance on sport hunting has raised the debate about whether sport hunting can be considered a form of tourism, and particularly whether it can be regarded as a new form of ethical tourism. CAMPFIRE clearly focuses on rural development and human needs as the priorities over notions of animal welfare. Although the wildlife authorities have been sensitive about charges of cruelty, the Parks Department is not against any form of wildlife utilization 'provided it falls within Zimbabwean Society's accepted norms of animal treatment' (Ministry of Environment and Tourism, 1992: 13). Critics of CAMPFIRE have been caught in a complicated ethical conundrum where those who argue against CAMPFIRE in terms of environmental ethics or animal welfare are presented as misanthropes who seek to deny rural people the right to use the resources around them. The difficulty is that the more we commodify the environment, the less scope there is for establishing ethical relations with the world around us and the more certain it is that human utility and global capitalism will become the sole arbiters of nature's uncertain future.

Conclusion

If nothing else, the cases of ecotourism in Belize and CAMPFIRE in Zimbabwe illustrate the enormous complexity of the ethical issues facing tourism development and the impossibility of finding a simple solution that will satisfy all concerned. They also demonstrate the multifarious ways in which economic, ethical and aesthetic values interact – sometimes clashing, sometimes combining to determine the trajectory of future developments. These value conflicts cannot always be resolved by appeal to the ethical frameworks we have examined in this book, but, as we argued in the Introduction, these discourses help frame and express particular forms of ethical concerns, about individual integrity, distributive justice, our duties to others, and so on. Philosophical theories about human or animal rights, utilitarianism or the virtues are not (whatever they may claim) neutral in these debates. They give voice to some aspects of our social and natural environment while repressing others, thereby helping to structure and give shape to a 'moral field'. In other words, ethical theories help both to create and to articulate the boundaries of moral debates by expressing heartfelt concerns for other individuals, for traditions and cultures, for our environment and, not least, for our future on this incredible planet. All development implies change, for the better and/or for the worse. Modernity feeds off such change often without considering its full implications. Economic growth in particular is seen as an unmitigated good, with the social, cultural and environmental impoverishment that have often followed in its wake being forgotten. If there is to be a world that is worth travelling to see, then we need to recognize and take steps to change this. The multiple languages of ethics are, one might say, ways in which we strive to *sustain* our concerns for those things that make life worth living in the face of change. In that case, genuine sustainable development is always and everywhere all about ethics.

8   Conclusion

We began this book by arguing that ethical values play an important but frequently overlooked part in controversies surrounding tourism development. This is true whether discussing the general idea of 'development' with all its political and economic connotations, particular forms of tourism – like sustainable tourism – or the specific inter-relations arising within individual resorts, tours, and so on. Ethics is, it seems, everywhere, and yet it remains difficult to articulate ethical values within a society and an industry dominated by commodity relations and where development means economic development (see Plate 8.1).

The very fact that ethical values (as we suggested in Chapter 1) remain surprisingly resistant to commodification means that they obtain little purchase in

Plate 8.1 Independence Day in Cuzco, Peru, 2000. The banner on the military bulldozer says 'Peru: A Country with a Future'.

commercially motivated developments. This, however, must change if tourism is ever to be sustainable in anything other than purely economic terms. While there is considerable disagreement about just what tourism should be sustaining (McCool *et al.*, 2001) there is general agreement that, as Vandana Shiva (1989) points out, it is not simply a matter of being able to carry on regardless into the distant future or extracting profits indefinitely. To sustain also means to 'bear up', to care for and consider the needs of others. Caring for, being considerate of, and sustaining are all ethical relations.

How, though, are we to address the obvious incompatibilities between the various frameworks we have presented for thinking through and expressing ethics? Utilitarianism and rights theories both emphasize important aspects of ethical relations to others, the feeling that happiness matters or that we deserve to have our freedom and life respected no matter what. But when push comes to shove and we try to apply them consistently or rigidly, then it becomes obvious that they often lead to anomalous conclusions. Sometimes they straightforwardly conflict with each other and with other frameworks like virtue theories (Chapter 2), an ethics of care or difference ethics (see Chapter 5). This does not mean that such theories and the language they deploy are useless or wrong. It simply points up the very real complexities inherent in the ethical relations they attempt to express and control. Ethical values are as complex as the societies within which they reside, and, as Durkheim argued (see Chapter 2), the extensive division of tasks in late modern society means that no single set of principles is likely to cover all eventualities. These different ways of speaking of values at least do something to make ethics tangible, to give it a *currency* when it comes to taking more than money into account.

It might be worth taking this notion of 'currency' further because in one sense these dialogues about ethics become especially necessary where a purely economic currency threatens to dominate social relations and social development – that is, precisely when society comes to be dominated by commodity relations. Resistance to, or amelioration of, the over-extension of market relations may seem difficult when even a traditionally anti-materialist religion like Buddhism can become commodified as a money-spinning tourist attraction (Philip and Mercer, 1999). Money is such an effective currency – indeed, the archetypal currency – because it can stand in for so many different commodities and because it can freely circulate among people in a tangible form. Indeed, the key point about a commodity is that it has an exchange-value that can be expressed in monetary terms, a monetary value which comes to be regarded as its *real* value, its worth in euros or dollars (see Chapter 1). The value of a holiday as a commodity is the amount of money that will be paid (exchanged) for it. This obviously means that we are putting a price on things like a flight, a hotel room, a suntan, a week's relaxation or a sea view. This seems quite normal to us and we are used to making the complex decisions that allow us to make comparisons between commodities, judgements about value for money, and so on. But sometimes we might, knowingly or unknowingly, be putting a price on other things, on other lives, cultures and environments – even on our own conscience.

Emphasizing exchange-value to the exclusion of all else necessarily underplays those aspects of things that are not exchangeable. Indeed, it obscures the very qualities that, somewhere along the line, form the basis of that thing having any value at all. Holidays, for example, provide us with many varied and valuable experiences: the beauty of a setting sun, the taste of an exotic meal, new friendships, a different way of seeing the world, a feeling of freedom, and so on. It is in these, and not in the monetary value itself, that the real value (of travel) lies. Imagine what you would be missing if all you could remember about a holiday was its cost! Yet in modern capitalist society the commodity form takes on a mysterious life of its own and comes to dominate the way in which we experience the world and our social relations. Money becomes the measure of all things, substituting itself for them as though exchange-value were all that mattered – as though its value had some mysterious origin and power outside of the things themselves.[1]

When it succeeds in this sleight of hand, then we risk slipping into a one-dimensional world ruled by commodity exchange. Here development just means maximizing monetary gains, and sustainability loses all those other aspects that Shiva believes make it so necessary and worthwhile. When it is only money that matters, then cultures, environments and people themselves are reduced to commodities, things that have only exchange-value, whose 'real' worth is not what they are in themselves but how much they can be traded for. This is truly the end of ethics since it means that we see people, places and things as nothing more than a replaceable means to financial enrichment. In respect of Kant's position (Chapter 4) we regard them as (instrumentally valuable) 'means' rather than (ethically valuable) 'ends' in themselves.

This is why tourism development, even when motivated by the best intentions of creating and distributing wealth, can sometimes be so destructive, replacing what is unique or irreplaceable with what is reproducible and saleable. And as this system of commodity production accelerates and expands, so values circulate faster and faster, becoming ever more detached from their place of origin, those specific contexts in which all substantial values originate. Such flows are not, of course, limited to money and commodities; they also include information, images and people themselves. So the late modern or

> postmodern political economy is one of the ever more rapid circulation of subjects and objects. But it is also one of the 'emptying out' of objects . . . time and space 'empty out', become more abstract; and in which things and people become 'disembedded'.
>
> (Lash and Urry, 1994: 13)

Tourism development becomes a literal and figurative process of disembedding subject and objects, images and experiences, of making it ever more difficult to trace things and values back to their place of origin, those contexts of particular social relations within which they first became significant. *Significance* is replaced by the *sign*, a £ or $ or €, and eventually these signs seem to form an evaluative reality of their own, a hyper-reality where it is impossible to trace the link between sign and 'true' significance. This world is that of the simulacrum, the sign without external signification (Baudrillard, 1994).

As the connection between place and value is severed, it becomes increasingly difficult to express anything about values outside of the system of economic exchange. The 'real' (non-exchange) value of things exists only in a bounded set of social relations that have been over-written and undermined by their inclusion within an abstract system of constant circulation, a system where the very idea of non-exchange values has no currency. Tourism as an agent of post-modernity 'problematizes the relationship between representations and reality' (Lash and Urry, 1994: 272), between touristic evaluations of what is presented as a place to visit and places themselves. This is also why (as Chapter 6 showed) questions of authenticity become difficult to answer. The call for authenticity requires us to trace something back to its origins in a particular place and time.

On this reading, ethics is in difficulty because it attempts to return to the things-in-themselves, or perhaps more accurately to things in the full complexity of their inter-relations with us. Ethics is a relation of genuine, heartfelt *significance*, not an exchange relation between signs or simulacra. It is, of course, not the only way of reminding us of what is really important, what stands as an alternative to and underlies exchange values. Marx, for example, chose to remind us of the use-value of things, the properties of things that make them useful to us. Others might emphasize the aesthetic qualities and experiences derived from others. But only ethics chooses to remind us of the importance of values beyond exchange, use or aesthetics – the value of things (people, cultures, environments) as ends-in-themselves.

The idea of an ethical tourism, a tourism that is environmentally and culturally sustainable in the full sense of the term, might, then, seem like a contradiction in terms. If tourism is all about mobility, circulation, replaceability – if it helps increase commodification and extend market relations – then it would seem to undermine any kind of ethics that might originate in and concern respect for certain place-specific values. As Campbell and Shapiro (1999: ix) state, 'all the world over we seem to be losing touch with those "moral spaces", the bounded locations whose inhabitants acquire the privileges deriving from practices of ethical inclusion'. Somehow an ethical tourism would have to modulate between the contrary processes leading to values becoming disembedded from and (re-)embedded in specific locations – namely, tourist destinations and the homelands of the 'guests'.

And here another problem arises, because ethical values are usually strongest where the relationship between self and other is maintained and developed over time and where self and other can be (emotionally and geographically) *close* to each other. Yet from the tourist's point of view, holidays entail only fleeting visits to distant places where personal or close contact with others is usually extremely limited. The possibilities for developing new ethical relations seem equally restricted, though the kind of community-based tourism projects discussed in Chapters 6 and 7 are possible ways to overcome the restrictions. Alternatively, we can envisage ways of generalizing and expanding the ethical relations we have with those close to us to others in similar positions but with whom our contact is more diffuse. This is, as Chapters 3 and 4 illustrated, precisely what discourses of rights and of utility attempt to do. They try to extrapolate from the particular and the personal to the universal and institutional in order to maintain ethical relations over the kind of distances that modernity requires. This is the source of both their power and their problems. They are able to exert influence because the relations between signs and significance, people and places, are not (yet) as anarchic and unstructured as theorists like Baudrillard believe. As Lash and Urry (1994: 12) point out, 'this flow of subjects and objects is not as free, not as "deregulated", as it might seem. Indeed the flows are highly specific to particular times and particular spaces.' In other words, rights and utility can operate as discourses within these regulatory and institutional frameworks to ensure that ethics has some currency.

But as we saw (in Chapters 3 and 4), such discourses are sometimes problematic because in order to achieve universality they have to effectively abstract themselves from specific contexts. In doing so there is the ever-present danger that they too will come to reflect the very processes they are meant to overcome, that they too become incapable of expressing the real ethical aspirations of different people. If, as Alasdair MacIntyre (1981: 4) argues, the moral crisis of modernity is 'so pervasive that it has invaded every aspect of our intellectual and moral lives, then what we take to be resources for the treatment of our condition may turn out themselves to be infected areas'. In other words, discourses of rights and utility, and also abstract notions of distributive justice, including Rawls's early formulations (Chapter 5), might also become too divorced from the specific contexts – the particular places and social relations – that are the origins of the values they try to express and enforce. When this happens, they too begin to privilege the *sign* of utility, rights or justice over the *significance* of actual ethical relations.

If these thoughts offer a plausible explanation of our contemporary condition, then a genuine ethical tourism might seem almost impossible. But while we hope the practical evidence we have provided gives some credence to such speculations, and while a healthy scepticism of much ethical theory is certainly necessary, things may not be quite so bleak. To be sure, the development of tourism has problematized the question of an authentic or genuine ethical relation to the people, cultures and environments we visit. But this is not something unique to tourism; it is a feature of late modernism itself, of a kind of society of which tourism is just a part, albeit an increasingly important part. If, as we suggested, ethics might be seen as an attempt to reach that which is significant about our non-instrumental relations to others, to recover and sustain the other as a thing-in-itself, then the modernity might seem to place the whole of morality into crisis. For given the constantly changing parameters of our lives, how can we ever know what someone else is really like, what they really need from us, what a genuine ethical response should really entail? Can we ever, as the ethics of care (Chapter 5) suggests, know how we should care for others? Can we ever recognize and respect the Other for their differences from us?

The problem of ethical 'authenticity', then, goes far deeper than simply whether or not a performance presented to the paying tourist is 'genuine' or not. It concerns whether the whole of late modern life, epitomized in a fairly extreme form by tourism, is no more than a commodified simulation where authenticity (and therefore ethics) is simply unattainable. This at least seems to be the case so long as we are looking for some kind of completion in our relations to others – that is, some form of practice, theory or code that will distil the genuine spirit of the other and of ethics from the confusing and corrosive flows of what Bauman (2000) terms 'liquid modernity'. The problem is that treating the other as an *end*-in-itself suggests that such completion is possible, that we can know the other completely, authentically. This belief in our ability to attain absolute knowledge is perhaps a vestige of an earlier, more optimistic form of modernity when our relation to others and the world around us seemed less ambiguous and more clear-cut. Today we also have less faith in our abilities to reason our way to absolute and timeless conclusions. We are more sceptical about the possibility of a full understanding of the Other, or even of ourselves.

To put it another way, in terms that Kant would recognize (see Chapter 4), there is a largely unrecognized, or at least suppressed, tension between epistemology (what we can know) and ethics, between never being able to know the thing-in-itself and yet striving to treat something as an end-in-itself. This tension generates the problems we have about authentic ethical relations, and this general problem arises in extreme forms in tourist relations. How can we know what the real role of Maya basket weaving (Chapter 6) or of elephant hunting in Zimbabwe (Chapter 7) may have been when our very presence distorts that relation? How, then, can we judge something authentic or know how we should respond ethically if we don't know what 'end' it is we should be striving to reach?

Perhaps, then, we should stop contrasting that instrumental approach which dominates contemporary social relations – treating the other as a *means* to our ends – with that of treating the other as an *end*-in-itself. Perhaps we need to recognize that ethics has always been more complex than this and that completion (authenticity), or the full expression of ethical significance with a system of signs, may always have been unreachable ideals. This is not just because of the difficulties in developing relations with and understanding others, though this is certainly important, but because others are *never* self-contained either; they are never simply *ends*-in-themselves. They too are not completed projects but beings in a constant process of *becoming* someone or something different. Other people do not have a fixed self-identity any more than we do. They age, alter, encounter and encompass new experiences. Other cultures and environments undergo similar changes. The problem is that tourism developments often impose these changes on them through external pressures, without choice, and at a speed that makes them destructive rather than constructive. This is, as we have argued throughout, the antithesis of an ethical relation. Sustainable tourism development certainly cannot be about treating others as if they should be preserved in aspic, but it must embody an enlightened approach to encountering and understanding the Other.

How is this understanding to be achieved? This is clearly not the place to develop yet another ethical theory or provide a simple answer. From the very start of this book we made it clear that this was a project that did not seek completion, that would not attempt to provide universal answers about which theories best articulated ethical issues. Instead, we have tried to understand where those theories most intimately connected with late modernity have come from and something of the limits of their application in the context of tourism development. We also hope that this book will open up a new and challenging area of debate. But despite the difficulties involved, we have also tried to show that there are at least some grounds for hope. The fact that there are no easy answers should not surprise or intimidate us. Ethical understanding is possible. We see it in our everyday lives, and the thousands of tourists, businesses and communities that work to make tourism more ethical point to the felt need we have for such relations. The few examples we have managed to detail in this book point to the imaginative diversity of the solutions on offer, and it is precisely in terms of diversity and imagination that hope lies.

The key problem of ethics seems to be that of understanding the Other. But all 'understanding', according to the philosopher Hans-Georg Gadamer, involves an imaginative attempt to see beyond the limited *horizons* (boundaries) of our current thoughts and feelings. Understanding arises only through the meeting of different horizons and in the imaginative transposition of ourselves into others' places. It involves attempting to interpret that which is other than ourselves by thinking beyond where we come from and seeing the world from another('s) point of view. This relation cannot be forced on the Other but requires us to develop a tactful response to their presence, one that both helps us understand their difference to us and preserves a respectful distance. 'It avoids the offensive, the intrusive, the violation of the intimate sphere of the other' (Gadamer, 1998: 16). In this sense Gadamer offers us both an interpretation of ethics and an ethics of interpretation, a way of relating to others that, in this sense at least, is not dissimilar to the difference ethics of Irigaray and Levinas (see Chapter 5). It also has close ties to the notion of ethical work we discussed in Chapter 3. We have to work upon ourselves to cultivate openness to others, to cultivate an active imagination.

Gadamer cannot offer us an ethical panacea, but this approach, based in his work, might help us recuperate a hope that varieties of ethical tourism can develop. Perhaps we never have an entirely authentic experience of the Other, but this does not mean that we cannot try to interpret their needs through developing an open ethical relation with them. Perhaps we cannot simply legislate for ethical tourism, but surely all legislation should be informed by the actual contexts of our ethical encounters with each other. This kind of responsive ethics can be cultivated in many different ways, and in one sense there is nowhere better to attempt this than in the field of tourism development. After all, what other experiences do we have that compare with travel as opportunities to expand our horizons (geographically, culturally, emotionally, intellectually), to encounter people, cultures and places so very different from ourselves?

Notes

1 Ethical values

1 This illustrates how the spread of global tourism is often contiguous with the spread of global capitalism. Marx argues that capitalism is 'characterized by the fact that labour-power, in the eyes of the worker himself, takes on the form of a commodity which is his property; his labour consequently takes on the form of wage labour' (1990: 274).
2 Durkheim criticized this tendency to completely separate economic and ethical questions: 'If the orthodox economists and moralists of the Kantian school place political economy outside of morality, it is because these two sciences seem to them to study two worlds with no connection between them. But . . . one understands nothing of the maxims of morality regarding property, contracts, labor etc., if one does not understand the economic causes from which they are derived. On the other hand, one would have a very false idea of economic development, if one were to ignore the moral causes that intervene in that development' (1993: 67).
3 The kind of philosophical approach we have in mind here is an attempt to make *logical* and absolute distinctions between moral and non-moral uses of terms like 'good', as in 'good behaviour' and 'good food' respectively (see, for example, Hare, 1952). There are many contexts where such absolute distinctions do not hold. For example, among people who genuinely believe in the claim 'you are what you eat', references to good food might well conjoin moral and non-moral claims: the food is of good quality and will make you morally a better person.
4 E.P. Thompson (1967) famously argued that industrial capitalism demands the technological conditioning of a workforce to accept time constraints and routines. A form of wage-labour synchronized through the 'tyranny' of clocks, watches, timetables, factory whistles, and so on replaces the task-oriented labour of pre-industrial societies in which work rhythms were responsive to the needs of the task in hand.
5 The term 'form of life' is used in a technical sense by the philosopher Ludwig Wittgenstein (1981) to denote the social and environmental 'background' that we call upon to understand a word's use and meaning (see also Davidson and Smith, 1999).
6 Although, writing when he did, Hobbes would not, of course, have used the term 'psychology'.
7 Hobbes goes on to argue that the laws of the land which regulate this competition and constrain its worst excesses are themselves a form of contractual obligation between individuals, enforced through the power they agree to cede to the monarch in return for peace.
8 Some of these issues are covered in work on the ways that the American managers at Disneyland Paris dealt with their French staff (Warren, 1999), and in the gendered dimensions of tourism development in Stirling (Aitchison, 1999).

2 The virtues of travel and the virtuous traveller

1 In fact, Durkheim chose to study the religious/ethical views of Aboriginal Australian cultures, which are certainly not simple and cannot be regarded as primitive versions of modern Western societies.
2 This situation is somewhat different in Japan, but it is no accident that there the company is usually regarded as having an obligation to provide a job for life.

3 The greatest happiness is to travel?

1 Anonymous interviewee (Belize). Pattullo (1996) discusses the ways that international tour companies can insist that local operators change their tour itineraries, rebuild hotel complexes and accept very low rates per room, and even force them to do so.

4 Rights and codes of practice

1 However, it might be argued that Locke's supposedly objective account of our pre-social human nature is actually a reflection of his own social circumstances. It is not accidental that it expresses that ideology of 'possessive individualism' (Macpherson, 1979) which would come to provide the political underpinnings of an emergent modern society with its liberal ideas of democracy and capitalist ideals of a market-based society.
2 Eduardo Galeano (1973) notes the CIA involvement in the overthrow of Cheddi Jaggan's socialist government in Guyana in 1964 in order to maintain a cheap supply of bauxite for Alcoa, the Aluminum Company of America. This overthrow was then supported by British troops. O'Shaughnessy (2000: 30) quotes from a report from Edward Korry (the US ambassador to Santiago at the time of Salvador Allende's election victory) to Henry Kissinger: 'Once Allende comes to power we shall do all within our power to condemn Chile and the Chileans to utmost deprivation and poverty.' As John Pilger notes, some 200,000 people on East Timor (about a third of the population) died at the hands of the US-trained and -backed Indonesian military. The US ambassador to the United Nations, Daniel Patrick Moynihan, claimed that 'the United States wished things to turn out as they did and worked to bring this about' (Pilger, 1992: 349).

5 From social justice to an ethics of care

1 Although this parallel should not be taken too far since dependency theorists regard dependency itself as something negative, foisted on the non-Western world, while relations of dependency, or at least inter-dependency, are generally regarded positively by Gilligan.

6 Authenticity and the ethics of tourism

1 Rosaleen Duffy carried out fifty-three semi-structured interviews during two periods of fieldwork from 1997 to 1999. These interviews were mostly with individual tourists, but occasionally with couples and groups of friends travelling together. As well as tourists, key representatives from the tourism industry were also interviewed. A total of eighty-nine semi-structured interviews were carried out (from 1997 to 2000) with tour operators, guides, academics, journalists and representatives from conservation organizations, community organizations and government departments. The interviews were individually tailored to the interviewee's specific area of expertise.
2 Advert for Manta Resort in *Destination Belize* (Belize Tourism Board, 1997).

8 Conclusion

1 This is, of course, what Marx referred to as commodity fetishism.

Bibliography

Abbink, J. (2000) 'Tourism and its discontents: Suri–tourist encounters in southern Ethiopia', *Social Anthropology*, 8: 1–17.

Abbott-Cone, C. (1995) 'Crafting selves: the lives of two Mayan women', *Annals of Tourism Research*, 22: 314–327.

Abrahamsen, R. (2001) *Disciplining Democracy: Development Discourse and Good Governance in Africa*, London: Zed Books.

Abrahamsen, R. and Williams, P. (2001) 'Ethics and foreign policy: the antinomies of New Labour's "Third Way" in sub-Saharan Africa', *Political Studies*, 49: 249–264.

Afshar, H. (1998) *Islam and Feminisms: An Iranian Case Study*, London: Macmillan.

Aitchison, C. (1999) 'Heritage and nationalism: gender and the performance of power', in D. Crouch (ed.) *Leisure/Tourism Geographies: Practices and Geographical Knowledge*, London: Routledge.

—— (2001) 'Theorizing other discourses of tourism, gender and culture: can the subaltern speak (in tourism)?', *Tourist Studies*, 1: 133–147.

Ajami, R.A. (1988) 'Strategies for tourism transnationals in Belize', *Annals of Tourism Research*, 15: 517–530.

Akama, J.S., Lant, C.L. and Wesley, G. (1996) 'A political ecology approach to wildlife conservation in Kenya', *Environmental Values*, 5: 335–347.

Alexander, J., McGregor, J. and Ranger, T. (2000) *Violence and Memory: One Hundred Years of the Dark Forests of Matabeleland*, Oxford: James Currey.

Allen, T. and Thomas, A. (eds) (1992) *Poverty and Development in the 1990s*, Milton Keynes: Open University Press.

Amandala (1997a) 'TEA replies to Toledo tour guides' – letter from TEA Executive Board, 13 April.

—— (1997b) 'Toledo Ecotourism Association wins TODO 96 Award', 20 April.

—— (1997c) 'Airlines in bed together: Tourism Minister Henry Young', 18 May.

—— (1997d) 'Toledo Tour Guide Association vexed with Ministry of Tourism' – letter from the Toledo Tour Guide Association, 22 June.

—— (1998) 'Mayan student irate' – letter from Valentino Shal, 26 September.

—— (1999a) 'Guatemalan forestry officials contemplate charges against Yalbac ranch', 13 June.

—— (1999b) Statement by Minister Jorge Espat on June 12 1999 border incident, 4 July.

—— (1999c) Press release regarding the report on the Malaysian logging review committee, 28 March.

—— (1999d) '17 testify in case of Guate shot by BDF', 1 August.

—— (1999e) 'Northern Maya-Mestizos support Valentino Shal's protest', 17 October.

—— (1999f) 'Territorial uncertainty', 17 October.

—— (1999g) 'Present state of affairs of Toledo Mayas', 21 November.

—— (1999h) 'Toledo's one room hotels in financial trouble', 21 November.

—— (1999i) 'Toledo Maya comment on Bilal Morris article' – letter from Pio Coc, 28 November.

Amsden, A. (1990) 'Third World industrialization: "global Fordism" or a new model?', *New Left Review*, 182: 5–32.

Anderson, B. (1991) *Imagined Communities: Reflections on the Origin and Spread of Nationalism*, London: Verso.

Ap, J. (1992) 'Residents' perceptions on tourism impacts', *Annals of Tourism Research*, 19: 665–690.

Apostopoulos, A., Leivadi, A. and Yiannakis, A. (eds) (1996) *The Sociology of Tourism*, London: Routledge.

Aramberri, J. (2001) 'The host should get lost: paradigms in tourism theory', *Annals of Tourism Research*, 28: 738–761.

Arendt, H. (1989) *Lectures on Kant's Political Philosophy*, Chicago: University of Chicago Press.

Ariel de Vidas, A. (1995) 'Textiles, memory and the souvenir industry in the Andes', in M.F. Lanfant, J.B. Allcock and E.M. Bruner (eds) *International Tourism: Identity and Change*, London: Sage.

Aristotle (1986) *Ethics*, Harmondsworth: Penguin.

Ateljevic, I. and Doorne, S. (2002) 'Representing New Zealand: tourism imagery and ideology', *Annals of Tourism Research*, 29: 648–667.

Bakhtin, M. (1994) 'Carnival ambivalence', in P. Morris (ed.) *The Bakhtin Reader: Selected Writings of Bakhtin, Medvedev, Voloshinov*, London: Arnold.

Barbier, E., Burgess, J., Swanson, T. and Pearce, D. (1990) *Elephants, Economics and Ivory*, London: Earthscan.

Barbier, E., Burgess, J.C. and Folke, C. (1994) *Paradise Lost? The Ecological Economics of Biodiversity*, London: Earthscan.

Barclay, J. and Ferguson, D. (1992) 'Caught between the devil and the deep blue sea: the development of tourism in Cuba', *Community Development Journal*, 27: 378–385.

Battersby, C. (1978) 'Morality and the Ik', *Philosophy*, 53: 201–224.

Baudrillard, J. (1993) *America*, London: Verso.

—— (1994) *Simulacra and Simulation*, Ann Arbor: University of Michigan Press.

Bauman, Z. (1993) *Post-modern Ethics*, Oxford: Blackwell.

—— (1995) *Life in Fragments: Essays in Post-modern Morality*, Oxford: Blackwell.

—— (1997) *Post-modernity and Its Discontents*, Cambridge: Polity.

—— (2000) *Liquid Modernity*, Cambridge: Polity.

Beckerman, W. (1994) 'Sustainable development: is it a useful concept?', *Environmental Values*, 3: 191–209.

Belize Tourism Board (1997) *Destination Belize 1997*, Belize City: Belize Tourism Board.

Benhabib, S. (1990) 'In the shadow of Aristotle and Hegel: communicative ethics and current controversies in practical philosophy', in M. Kelly (ed.) *Hermeneutics and Critical Theory in Ethics and Politics*, Cambridge, MA: MIT Press.

Bentham, J. (1907) *An Introduction to the Principles of Morals and Legislation*, Oxford: Oxford University Press.

—— (1948) 'A fragment on government', in W. Harrison (ed.) *A Fragment on Government and An Introduction to the Principles of Morals and Legislation*, Oxford: Blackwell.

—— (1987) 'Introduction to the *Principles of Morals and Legislation*', in A. Ryan (ed.) *Utilitarianism and Other Essays: J.S. Mill and Jeremy Bentham*, Harmondsworth, UK: Penguin.

Benz, S. (1998) *Green Dreams: Travels in Central America*, Melbourne: Lonely Planet Publications.

Bernstein, J.M. (1992) *The Fate of Art: Aesthetic Alienation from Kant to Derrida and Adorno*, Cambridge: Polity Press.

Besculides, A., Lee, M.E. and McCormick, P.J. (2002) 'Residents' perceptions of the cultural benefits of tourism', *Annals of Tourism Research*, 29: 303–319.

BETA (1999) Belize Eco-Tourism Association Code of Ethics. Online. Available at: <http://www.belizenet.com/beta/ethics.html> (accessed 9 October 2000).

Biersteker, T. (1995) 'The "triumph" of liberal economic ideas in the developing world', in B. Stallings (ed.) *Global Change, Regional Response*, Cambridge: Cambridge University Press.

Birkeland, I.J. (2002) *Stories from the North: Travel as Place-Making in the Context of Modern Holiday Travel to North Cape, Norway*, Oslo: Institutt for Sosiologi og Samfunnsgeografi.

Bloom, J. (1996) 'A South African perspective of the effects of crime and violence on the tourism industry', in A. Pizam and Y. Mansfeld (eds) *Tourism, Crime and International Security Issues*, Chichester: John Wiley.

Bolland, O.N. (1997) *Struggles for Freedom: Essays on Slavery, Colonialism and Culture in the Caribbean and Central America*, Belize City: Angelus Press.

Bonner, R. (1993) *At the Hand of Man: Peril and Hope for Africa's Wildlife*, London: Simon and Schuster.

Boo, E. (1990) *Ecotourism: The Potentials and the Pitfalls*, vols 1 and 2, Washington, DC: WWF.

Bottrill, C.G. (1995) 'Ecotourism: towards a key elements approach to operationalizing the concept', *Journal of Sustainable Tourism*, 3: 45–54.

Bourdieu, P. (1998) *Distinction: A Social Critique of the Judgement of Taste*, London: Routledge.

Bramwell, B. and Lane, B. (1993) 'Interpretation and sustainable tourism: the potential and the pitfalls', *Journal of Sustainable Tourism*, 1: 71–80.

Brockington, D. (2002) *Fortress Conservation: The Preservation of the Mkomazi Game Reserve, Tanzania*, Oxford: James Currey.

Brohman, J. (1996) 'New directions in tourism for Third World development', *Annals of Tourism Research*, 23: 48–70.

Brown, D. (1996) 'Genuine fakes', in T. Selwyn (ed.) *The Tourist Image: Myths and Myth Making in Tourism*, Chichester, UK: John Wiley.

Brown, N. (1992) 'Beachboys as culture brokers in Bakau Town, The Gambia', *Community Development Journal*, 27: 361–370.

Brownlie, I. (1994) *Basic Documents in Human Rights*, Oxford: Oxford University Press.

Bruntland, G.H. (1987) *Our Common Future*, World Commission on Environment and Development, Oxford: Oxford University Press.

Bunn, S. (2000) 'Stealing souls for souvenirs: or why tourists want the "real thing"', in M. Hitchcock and K. Teague (eds) *Souvenirs: The Material Culture of Tourism*, Aldershot, UK: Ashgate.

Burns, P. (1999) 'Paradoxes in planning: tourism elitism or brutalism?', *Annals of Tourism Research*, 26: 329–348.

Butcher, J. (2002) *Debating Matters – Ethical Tourism: Who Benefits?* London: Hodder and Stoughton.

Butler, R. and Hinch, T. (eds) (1996) *Tourism and Indigenous Peoples*, London: International Thomson Business Press.

Campbell, D. and Shapiro, M.J. (1999) *Moral Spaces: Rethinking Ethics and World Politics*, Minneapolis: University of Minnesota Press.

Care for the Wild (1992) *The Elephant Harvest? An Ethical Approach*, Rusper, West Sussex, UK: CFTW.

Carr, A.Z. (1968) 'Is business bluffing ethical?', *Harvard Business Review*, January–February: 142–152.

Cater, E. (1992) 'Profits from paradise', *Geographical Magazine*, 64: 16–21.

—— (1994) 'Ecotourism in the Third World: problems and prospects for sustainability', in E. Cater and G. Lowman (eds) *Ecotourism: A Sustainable Option*, Chichester, UK: John Wiley.

—— (1995) 'Environmental contradictions in sustainable tourism', *Geographical Journal*, 161: 21–28.

Cater, E. and Lowman, G. (eds) (1994) *Ecotourism: A Sustainable Option*, Chichester, UK: John Wiley.

Chambers, R. (1983) *Rural Development: Putting the Last First*, Harlow, UK: Longman.

Chandler, D. (2002) *From Kosovo to Kabul: Human Rights and International Intervention*, London: Pluto Press.

Chant, S. (1992) 'Tourism in Latin America: perspectives from Mexico and Costa Rica', in D. Harrison (ed.) *Tourism and the Less Developed Countries*, London: Belhaven Press.

Child, B. (1988) 'The role of wildlife utilization in the sustainable economic development of semi arid rangelands in Zimbabwe', unpublished thesis, Oxford University.

—— (1995a) 'Can devolved management conserve and develop the management of natural resources in marginal rural economies? The example of CAMPFIRE in Zimbabwe', unpublished paper, Harare: DNPWLM.

—— (1995b) 'Guidelines for the revenue distribution process', unpublished policy draft for CAMPFIRE, Harare: DNPWLM.

—— (1995c) 'Guidelines for managing wildlife revenues in Communal Lands in accordance with policy for wildlife', unpublished policy draft, Harare: DNPWLM.

—— (1995d) *Communal Land Quotas for 1995*, Harare: DNPWLM.

Chodorow, N. (1978) *The Reproduction of Mothering*, Berkeley: University of California Press.

Ciulla, J.B. (1991) 'Why is business talking about ethics? Reflections on foreign conversations', *California Management Review*, Fall: 67–85.

Clancy, M.J. (1999) 'Tourism and development: evidence from Mexico', *Annals of Tourism Research*, 26: 1–20.

Cohen, E. (1972) 'Towards a sociology of international tourism', *Social Research*, 39: 164–182.

—— (1979) 'A phenomenology of tourist experiences', *Sociology*, 13: 179–201.

—— (1987) 'Alternative tourism: a critique', *Tourism Recreation Research*, 12(2): 13–18.

—— (1988) 'Authenticity and commodification in tourism', *Annals of Tourism Research*, 15: 371–386.

—— (1992) 'Pilgrimage centres: concentric and excentric?', *Annals of Tourism Research*, 19: 33–50.

Cohen, J. (2001) 'Textile, tourism and community development', *Annals of Tourism Research*, 28: 378–398.

Cohen, J.M. and Cohen, M.J. (1960) *The Penguin Dictionary of Quotations*, Harmondsworth, UK: Penguin.

Compagnon, D. (2000) 'Zimbabwe: Life after ZANU-PF', *African Affairs*, 99: 449–453.

Cothran, D.A. and Cole-Cothran, C. (1998) 'Promise or political risk for Mexican tourism', *Annals of Tourism Research*, 25: 477–497.

Craig, G.C. and Gibson, D.St.C. (1993) *Records of Elephant Hunting Trophies Exported from Zimbabwe*, Harare: DNPWLM.

Craik, J. (1995) 'Are there cultural limits to tourism?', *Journal of Sustainable Tourism*, 3: 87–98.

Crandall, R. (1980) 'Motivations for leisure', *Journal of Leisure Research*, 12: 45–54.

Crocker, D. (1991) 'Toward development ethics', *World Development*, 19: 457–483.

Crouch, D. (ed.) (1999) *Leisure/Tourism Geographies: Practices and Geographical Knowledge*, London: Routledge.

Cumming, D.H.M. (1990a) *Wildlife Conservation in African Parks: Progress, Problems and Prescriptions*, WWF MAPS Project Paper 5, Harare: WWF/DNPWLM.

—— (1990b) *Wildlife Products and the Marketplace: A View from Southern Africa*, WWF MAPS Project Paper 12, Harare: WWF.

Cumming, D.H.M. and Bond, I. (1991) *Animal Production in Southern Africa: Peasant Practice and Opportunities for Peasant Farmers in Arid Lands*, WWF MAPS Project Paper 22, Harare: WWF.

Cunliffe, R.N. (1994) *The Impact of the Ivory Ban on Illegal Hunting of Elephants in Zimbabwe*, WWF MAPS Project Paper 44, Harare: WWF.

Dahles, H. (2002) 'The politics of tour guiding: image management in Indonesia', *Annals of Tourism Research*, 29: 783–800.

Daily Gazette (1993) 'Locals Now Pay to Visit National Parks', 4 September.

Dann, G. (1996a) 'The people of tourist brochures', in T. Selwyn (ed.) *The Tourist Image: Myths and Myth Making in Tourism*, Chichester, UK: John Wiley.

—— (1996b) *The Language of Tourism: A Sociolinguistic Perspective*, Wallingford, UK: CAB International.

Davidson, J. (2001) 'Fear and trembling in the mall: women, agoraphobia and body boundaries', in I. Dyck, N.D. Lewis and S. McLafferty (eds) *Geographies of Women's Health*, London: Routledge.

Davidson, J. and Smith, M. (1999) 'Wittgenstein and Irigaray: philosophy in a language (game) of difference', *Hypatia: Journal of Feminist Philosophy*, 14(2): 76–96.

de Albuquerque, K. and McElroy, J.L. (2001) 'Tourist harassment: Barbados survey results', *Annals of Tourism Research*, 28: 477–492.

de Burlo, C. (1996) 'Cultural resistance and ethnic tourism on South Pentecost, Vanuatu', in R. Butler and T. Hinch (eds) *Tourism and Indigenous Peoples*, London: International Thomson Business Press.

De Waal, A. (1997) *Famine Crimes: Politics and the Disaster Relief Industry in Africa*, Oxford: James Currey.

Dent, K. (1975) 'Travel as education: the English landed classes in the eighteenth century', *Educational Studies*, 1: 171–180.

Desforges, L. (2000) 'Travelling the world: identity and travel biography', *Annals of Tourism Research*, 27: 926–945.

Devall, B. and Sessions, G. (1985) *Deep Ecology: Living as Though Nature Mattered*, Salt Lake City: Gibbs Smith.

Dharmaratne, G., Yee Sang, F. and Walling, L.J. (2000) 'Tourism potentials for financing protected areas', *Annals of Tourism Research*, 27: 590–610.

Dicks, B. (2000) *Heritage, Place and Community*, Cardiff: University of Wales Press.

Dickson, B. (2000) 'The ethicist conception of environmental problems', *Environmental Values*, 9: 127–152.

Dieke, P. (1993) 'Tourism and development policy in The Gambia', *Annals of Tourism Research*, 20: 423–460.

—— (1994) 'The political economy of tourism in The Gambia', *Review of African Political Economy*, 21: 611–626.

Dive Guide International. Online: Available at: <http://www.diveguideint.com/p0875. html> (accessed 19 March 1999).

DNPWLM (1991) *Protected Species of Animals and Plants in Zimbabwe*, Harare: DNPWLM.

—— (1992a) *Research Plan: DNPWLM, Zimbabwe*, Harare: Branch of Aquatic Ecology and Branch of Terrestrial Ecology, Research Division, DNPWLM.

—— (1992b) *Short and Medium Term Action Plans for Black Rhino*, Harare: DNPWLM.

—— (1992c) *Zimbabwe Black Rhino Conservation Strategy*, Harare: DNPWLM.

—— (1994) *Summary Report on the CAMPFIRE Programme and the CAMPFIRE Co-ordination Unit*, Harare: DNPWLM.

Doherty, F. (1995) 'Pagan', *Tourism in Focus*, 15: 9.

Donnelley, J. (1999) 'Post cold-war reflections on the study of international human rights', in J.H. Rosenthal (ed.) *Ethics and International Affairs: A Reader*, Washington, DC: Georgetown University Press.

Douzinas, C. (2000) *The End of Human Rights*, Oxford: Hart.

Doxey, G.V. (1975) 'A causation theory of visitor–resident irritants: methodology and research inferences', in *Proceedings of the Travel Research Association 6th Annual Conference*, San Diego: Travel Research Association.

Drinkwater, M. (1991) *The State and Agrarian Change in Zimbabwe*, London: Macmillan.

Drumm, A. (1998) 'New approaches to community-based ecotourism management: learning from Equador', in K. Lindberg, M.E. Wood and D. Engeldrum (eds) *Ecotourism: A Guide for Planners and Managers*, vol. 2, North Bennington, VT: Ecotourism Society.

Dryzek, J. (1999) 'Global ecological democracy', in N. Low (ed.) *Global Ethics and the Environment*, London: Routledge.

Duffield, M. (2001) *Global Governance and the New Wars*, London: Zed Books.

Duffy, R. (2000a) 'Shadow players: ecotourism development, corruption and state politics in Belize', *Third World Quarterly*, 21: 549–565.

—— (2000b) *Killing for Conservation: Wildlife Policy in Zimbabwe*, Oxford: James Currey.

—— (2001) 'Peace parks: the paradox of globalization', *Geopolitics*, 6: 1–26.

—— (2002) *A Trip Too Far: Ecotourism, Politics and Exploitation*, London: Earthscan.

Durkheim, É. (1968) *The Elementary Forms of the Religious Life*, London: George Allen and Unwin.

—— (1993) *Ethics and the Sociology of Morals*, New York: Prometheus.

—— (1997) *The Division of Labor in Society*, New York: Free Press.

Eckersley, R. (1992) *Environmentalism and Political Theory: Towards an Ecocentric Approach*, London: UCL Press.

Edensor, T. (2000) 'Staging tourism: tourists as performers', *Annals of Tourism Research*, 27: 322–344.

Eder, K. (1996) *The Social Construction of Nature*, London: Sage.

Edgeworth, F.Y. (1881) *Mathematical Psychics*, London: Kegan Paul.

Edwards, E. (1996) 'Postcards: greetings from another world', in T. Selwyn (ed.) *The Tourist Image: Myths and Myth Making in Tourism*, Chichester, UK: John Wiley.

El Sadaawi, N. (1997) *Why Keep Asking Me about My Identity: The Nawal El Sadaawi Reader*, London: Zed Books.

Elsrud, T. (2001) 'Risk creation in travelling: backpacker adventure narration', *Annals of Tourism Research*, 28: 597–617.

Enloe, C. (1990) *Bananas, Beaches and Bases: Making Feminist Sense of International Politics*, Berkeley: University of California Press.

Escobar, A. (1995) *Encountering Development: The Making and Unmaking of the Third World*, Princeton, NJ: Princeton University Press.

ESTAP (2000) *Regional Development Plan for Southern Belize*, ESTAP/GOB/Inter-American Development Bank: Belize City.

Estrada-Claudio, S. (1992) 'Unequal exchanges: international tourism and overseas employment', *Community Development Journal*, 27: 401–410.

Evernden, N. (1992) *The Social Creation of Nature*, Baltimore: Johns Hopkins University Press.

Farrell, B.H. and Runyan, D. (1991) 'Ecology and tourism', *Annals of Tourism Research*, 20: 26–40.

Fennell, D.A. (1999) *Ecotourism: An Introduction*, London: Routledge.

Ferguson, J. and Chisholm, K. (1978) *Political and Social Life in the Great Age of Athens*, London: Open University Press.

Financial Gazette (1995a) 'Sport hunting quotas cut', 12 January.

——— (1995b) 'Tourists hit one million mark', 6 July.

Forsyth, T. (1995) 'Business attitudes to sustainable tourism: self-regulation in the UK outgoing tourism industry', *Journal of Sustainable Tourism*, 3: 210–231.

Foucault, M. (1996) *Foucault Live: Collected Interviews, 1961–1984*, ed. S. Lotringer, New York: Semiotext(e).

Fukuyama, F. (1992) *The End of History and the Last Man*, London: Hamilton.

Gadamer, H.G. (1998) *Truth and Method*, New York: Continuum.

Galani-Moutafi, V. (2000) 'The self and the Other: traveler, ethnographer, tourist', *Annals of Tourism Research*, 27: 203–224.

Galeano, E. (1973) *The Open Veins of Latin America: Five Centuries of the Pillage of a Continent*, New York: Monthly Review Press.

Garber, M., Hansesen, B. and Walkowitz, R.L. (2000) *The Turn to Ethics*, London: Routledge.

Garrod, B. and Fyall, A. (2000) 'Managing heritage tourism', *Annals of Tourism Research*, 27: 682–708.

Gehrels, B. (1997) 'Namibia, rhetoric or reality?', *Tourism in Focus*, 23: 15–16.

Gergan, K.J. (1991) *The Saturated Self: Dilemmas of Identity in Contemporary Life*, New York: Basic Books.

Gewirth, A. (1978) *Reason and Morality*, Chicago: University of Chicago Press.

Gibson, H. and Yiannakis, A. (2002) 'Tourist roles: needs and the lifecourse', *Annals of Tourism Research*, 29: 358–383.

Giddens, A. (1999) *Modernity and Self-Identity: Self and Society in the Late Modern Age*, Cambridge: Polity.

Gide, A. (1966) *The Immoralist*, Harmondsworth, UK: Penguin.

Gilligan, C. (1983) *In a Different Voice: Psychological Theory and Women's Development*, Cambridge, MA: Harvard University Press.

—— (1994) 'Reply to my critics', in M.J. Larabee (ed.) *An Ethic of Care: Feminist and Interdisciplinary Perspectives*, London: Routledge.

Gnoth, J. (1997) 'Tourism motivation and expectation formation', *Annals of Tourism Research*, 24: 283–304.

Godwin, W. (1976) *Enquiry concerning Political Justice*, Harmondsworth, UK: Pelican.

Gorringe, T. (1999) *Fair Shares: Ethics and the Global Economy*, London: Thames and Hudson.

Government of Belize (1995) 'Tourism consultation report' (draft), Belmopan: Government of Belize.

Graburn, N.N.H. (1983) 'Tourism and prostitution', *Annals of Tourism Research*, 10: 437–442.

—— (1989) 'Tourism: the sacred journey', in V.L. Smith (ed.) *Hosts and Guests: The Anthropology of Tourism*, Philadelphia: University of Pennsylvania Press.

Greenwood, D.J. (1989) 'Culture by the pound: an anthropological perspective on tourism as cultural commoditization', in V.L. Smith (ed.) *Hosts and Guests: The Anthropology of Tourism*, Philadelphia: University of Pennsylvania Press.

Habermas, J. (1990) *Moral Consciousness and Communicative Action*, Cambridge: Polity Press.

—— (1993) *Justification and Application: Remarks on Discourse Ethics*, Cambridge, MA: MIT Press.

—— (1995) *Between Facts and Norms: Contributions to a Discourse Theory of Law and Democracy*, Cambridge: Polity.

Hadot, P. (1995) *Philosophy as a Way of Life*, Oxford: Blackwell.

Hale, A. (2001) 'Representing the Cornish: contesting heritage interpretation in Cornwall', *Tourism Studies*, 1(2): 185–196.

Halewood, C. and Hannam, K. (2001) 'Viking heritage tourism: authenticity and commodification', *Annals of Tourism Research*, 28: 565–580.

Hall, C.M. (1994) *Tourism and Politics: Policy, Power and Place*, Chichester, UK: John Wiley.

—— (1997) 'Sex tourism in South-East Asia', in L. France (ed.) *The Earthscan Reader in Sustainable Tourism*, London: Earthscan.

Hall, C.M. and O'Sullivan, V. (1996) 'Tourism, political stability and violence', in A. Pizam and Y. Mansfeld (eds) *Tourism, Crime and International Security Issues*, Chichester, UK: John Wiley.

Hall, D.R. (1992) 'Tourism development in Cuba', in D. Harrison (ed.) *Tourism and the Less Developed Countries*, London: Belhaven Press.

Hampton, M.P. (1998) 'Backpacker tourism and economic development', *Annals of Tourism Research*, 25: 639–660.

Hand, S. (1989) *The Levinas Reader*, Oxford: Blackwell.

Handler, R. and Saxton, W. (1988) 'Dissimulation: reflexivity, narrative, and the quest for authenticity', *Cultural Anthropology*, 3: 242–260.

Handszuh, H. (1998) Global Code of Ethics in Tourism. Online. Available at: <http://www.mcb.co.uk/services/conferen/jan98/eit/1_handszuh.html> (accessed 9 October 2000).

Hardin, G. (1968) 'The tragedy of the commons', *Science*, 162: 1243–1248.

Hare, R.M. (1952) *The Language of Morals*, London: Routledge.

Hargrove, E. (1989) *Foundations of Environmental Ethics*, Denton, TX: Environmental Ethics Books.

Harrison, D. (ed.) (1992a) *Tourism and the Less Developed Countries*, London: Belhaven Press.

—— (1992b) 'International tourism and the less developed countries: the background', in D. Harrison (ed.) *Tourism and the Less Developed Countries*, London: Belhaven Press.

—— (1992c) 'Tourism to less developed countries: the social consequences', in D. Harrison (ed.) *Tourism and the Less Developed Countries*, London: Belhaven Press.

Harrisson, L.E. (1993) 'Underdevelopment is a state of mind', in M.A. Seligson and J.T. Passe-Smith (eds) *Development and Underdevelopment: The Political Economy of Inequality*, London: Lynne Rienner.

Harsanyi, J. (1982) 'Morality and the theory of rational behaviour', in A. Sen and B. Williams (eds) *Utilitarianism and Beyond*, Cambridge: Cambridge University Press.

Harvey, D. (1990) *The Condition of Postmodernity*, Oxford: Blackwell.

Healy, R.G. (1994) 'Tourist merchandise as a means of generating local benefits from ecotourism', *Journal of Sustainable Tourism*, 2: 137–151.

Heckman, S. (1995) *Moral Voices, Moral Selves: Carol Gilligan and Feminist Moral Theory*, Cambridge: Polity.

Held, V. (ed.) (1995) *Justice and Care: Essential Readings in Feminist Ethics*, Boulder, CO: Westview Press.

Herald (1989) 'Centenarians don't want to lag behind', 26 November.

—— (1993) 'Entry fees for locals at resorts', 4 September.

—— (1994) 'Allow trade in legally obtained ivory and rhino horns: CAMPFIRE', 16 December.

—— (1995) 'Explain discounts in parks entry fees', 11 February.

Herlihy, P.H. (1997) 'Indigenous peoples and the biosphere reserve conservation in the Mosquitia Rain Forest Corridor, Honduras', in S. Stevens (ed.) *Conservation through Cultural Survival: Indigenous Peoples and Protected Areas*, Washington, DC: Island Press.

Herold, E., Garcia, R. and DeMoya, T. (2001) 'Female tourists and beach boys: romance or sex tourism?', *Annals of Tourism Research*, 28(4): 978–997.

Hillery, M., Nancarrow, B., Griffin, G. and Syme, G. (2001) 'Tourist perception of environmental impact', *Annals of Tourism Research*, 28: 853–867.

Hitchcock, M., King, V.T. and Parnwell, M.J.R. (eds) (1993) *Tourism in South East Asia*, London: Routledge.

Hitchcock, M. and Teague, K. (eds) (2000) *Souvenirs: The Material Culture of Tourism*, Aldershot, UK: Ashgate.

Hobbes, T. (1960) *Leviathan, or the Matter, Forme, and Power of a Commonwealth Ecclesiastical and Civil*, Oxford: Blackwell.

Hobsbawm, E. and Ranger, T.O. (1983) *The Invention of Tradition*, Cambridge: Cambridge University Press.

Hochschild, A.R. (1983) *The Managed Heart: Commercialization of Human Feeling*, Berkeley: University of California Press.

Honey, M. (1999) *Ecotourism and Sustainable Development: Who Owns Paradise?*, Washington, DC: Island Press.

Hoogvelt, A. (2001) *Globalization and the Post-colonial World*, 2nd edn, London: Palgrave.

Hovinen, G.R. (2002) 'Revisiting the destination lifecycle model', *Annals of Tourism Research*, 29: 209–230.

Hughes, P. (2001) 'Animals, values and tourism: structural shifts in UK dolphin tourism provision', *Tourism Management*, 22: 321–329.

Hulme, D. and Murphree, M. (eds) (2001) *African Wildlife and Livelihoods*, London: Heinemann.

Hultsman, J. (1995) 'Just tourism: an ethical framework', *Annals of Tourism Research*, 22: 553–567.

Huntington, S.P. (1968) *Political Order in Changing Societies*, New Haven, CT: Yale University Press.

Hurungwe District CAMPFIRE Newsletter no. 5, December 1993.

International Association of Convention and Visitor Bureaus (2000) World Tourism Organization Code of Ethics. Online. Available at: <http://www.iacvb.org/wto. html> (accessed 9 October 2000).

Irigaray, L. (1993) *An Ethics of Sexual Difference*, London: Athlone.

Jahoda, G. (1999) *Images of Savages: Ancient Roots of Modern Prejudices in Western Culture*, London: Routledge.

Jameson, F. (1991) *Postmodernism or the Cultural Logic of Late Capitalism*, London: Verso.

Jansen, D.J. (undated) *What Is a Joint Venture? Guidelines for District Councils with Appropriate Authority*, WWF MAPS Project Paper 16, Harare: WWF.

Joas, H. (2000) *The Genesis of Values*, Oxford: Polity.

Johnson, V. and Nurick, R. (1995) 'Behind the headlines: the ethics of the population and environment debate', *International Affairs*, 71: 547–565.

Johnston, L. (2001) '(Other) bodies and tourism studies', *Annals of Tourism Research*, 28: 180–201.

Joseph, C.A. and Kavoori, A.P. (2001) 'Mediated resistance: tourism and the host community', *Annals of Tourism Research*, 28: 998–1009.

Kaltenborn, B., Haaland, H. and Sandell, K. (2001) 'The public right of access: some challenges to sustainable tourism development in Scandinavia', *Journal of Sustainable Tourism*, 9: 417–433.

Kant, I. (1996) *The Metaphysics of Morals*, Cambridge: Cambridge University Press.

Kaplan, C. (1996) *Questions of Travel*, Durham, NC: Duke University Press.

Kasere, S. (1995) 'CAMPFIRE: Zimbabwe's tradition of caring', *CAMPFIRE Association Publication Series*, 1: 6–17.

Katz, E., Light, A. and Rothenberg, D. (2000) *Beneath the Surface: Critical Essays in the Philosophy of Deep Ecology*, Cambridge, MA: MIT Press.

Keefe, J. (1995) 'Whose home is it anyway?', *Tourism in Focus*, 15: 4–16.

Kellert, S.R. and Wilson, E.O. (eds) (1993) *The Biophilia Hypothesis*, Washington, DC: Island Press.

Kousis, M. (2000) 'Tourism and the environment: a social movements perspective', *Annals of Tourism Research*, 27: 468–489.

Krauss, R.E. (1996) *The Optical Unconscious*, Cambridge, MA: MIT Press.

Krippendorf, J. (1982) 'Towards new tourism policies', *Tourism Management*, 3: 135–148.

—— (1987) *The Holiday Makers: Understanding the Impact of Leisure and Travel*, London: Heinemann.

—— (1997) 'The motives of the mobile leisureman: travel between norm, promise and hope', in L. France (ed.) *The Earthscan Reader in Sustainable Tourism*, London: Earthscan.

Lanfant, M.F. (1995) 'International tourism, internationalization and the challenge to identity', in M.F. Lanfant, J.B. Allcock and E.M. Bruner (eds) *International Tourism: Identity and Change*, London: Sage.

Lanfant, M.F., Allcock, J.B. and Bruner, E.M. (eds) (1995) *International Tourism: Identity and Change*, London: Sage.

Langman, L. (1994) 'Neon cages: shopping for subjectivity', in R. Shields (ed.) *Lifestyle Shopping*, London: Routledge.

Lash, S. and Urry, J. (1994) *Economies of Sign and Space*, London: Sage.

Latouche, S. (1993) *In the Wake of the Affluent Society: An Exploration of Post-development*, London: Zed Books.

Le Dœuff, M. (1990) *Hipparchia's Choice: An Essay concerning Women, Philosophy, etc.*, Oxford: Blackwell.

Lea, J.P. (1993) 'Tourism development ethics in the Third World', *Annals of Tourism Research*, 20: 710–715.

Leach, M. and Mearns, R. (eds) (1996) *The Lie of the Land: Challenging Received Wisdom on the African Environment*, Oxford: James Currey.

Leader-Williams, N. (1994) 'Sustainable use of elephants, with a focus on East Africa', paper presented at Conference on the African Elephant in the Context of CITES, Kasane, Botswana, 19–23 September 1994.

Lee, T.H. and Crompton, J. (1992) 'Measuring novelty seeking in tourism', *Annals of Tourism Research*, 19: 732–751.

Lehmann, D. (1997) 'An opportunity lost: Escobar's deconstruction of development', *Journal of Development Studies*, 33: 568–578.

Leopold, A. (1949) *A Sand County Almanac*, New York: Oxford University Press.

Levinas, E. (1991) *Totality and Infinity*, Dordrecht: Kluwer Academic.

Lewis, N. (1984) *Voices of the Old Sea*, London: Hamish Hamilton.

Li, Y. (2000) 'Ethnic tourism: a Canadian experience', *Annals of Tourism Research*, 27: 115–131.

Liebman Parrinello, G. (1993) 'Motivation and anticipation in post-industrial tourism', *Annals of Tourism Research*, 20: 233–249.

Lindberg, K. (1991) *Policies for Maximizing Nature: Tourism's Ecological and Economic Benefits*, New York: World Resources Institute.

Lindberg, K. and Hawkins, D.E. (eds) (1993) *Ecotourism: A Guide for Planners and Managers*, North Bennington, VT: Ecotourism Society.

Lindberg, K., Enriquez, K.J. and Sproule, K. (1996) 'Ecotourism questioned: case studies from Belize', *Annals of Tourism Research*, 23: 543–562.

Lindberg, K., Andersson, T.D. and Dellaert, B.G.C. (2001) 'Tourism development: assessing social gains and losses', *Annals of Tourism Research*, 28: 1010–1030.

Ling, C.Y. (1995) 'A rough deal: golf displaces people', *Tourism in Focus*, 15: 12–13.

Lingis, A. (1995) *Abuses*, Berkeley, CA: University of California Press.

Lloyd, G. (1984) *The Man of Reason: 'Male' and 'Female' in Western Philosophy*, London: Methuen.

Locke, J. (1988) 'An essay concerning the true, original extent and end of civil government', in P. Laslett (ed.) *Two Treatises of Government*, Cambridge: Cambridge University Press.

Loon, R.M. and Polakow, D. (2001) 'Ecotourism ventures: rags or riches?', *Annals of Tourism Research*, 28: 892–907.

Low, N. and Gleeson, B. (1999) 'Geography, justice and the limits of rights', in J.D. Proctor and D.M. Smith (eds) *Geography and Ethics: Journeys in a Moral Terrain*, London: Routledge.

Lukes, S. (1988) *Émile Durkheim – His Life and Work: A Historical Overview*, London: Penguin.

—— (1995) *The Curious Enlightenment of Professor Caritas: A Novel*, London: Verso.

Lynn, W. (1992) 'Tourism in the people's interest', *Community Development Journal*, 27: 371–377.

Lyons, D. (1965) *Forms and Limits of Utilitarianism*, Oxford: Oxford University Press.

Lyotard, J.-F. (1984) *The Post-modern Condition: A Report on Knowledge*, Manchester: Manchester University Press.

McAllister, D. (1980) *Evaluation in Environmental Planning*, Cambridge, MA: MIT Press.

MacCannell, D. (1999) *The Tourist: A New Theory of the Leisure Class*, 3rd edn, Berkeley, CA: University of California Press.

McCarthy, T. (1984) *The Critical Theory of Jürgen Habermas*, Cambridge: Polity.

—— (1993) *Ideals and Illusions: On Reconstruction and Deconstruction in Contemporary Moral Theory*, Cambridge, MA: MIT Press.

McCool, S.F., Moisey, R.N. and Nickerson, N.P. (2001) 'What should tourism sustain? The disconnect with industry perceptions of useful indicators', *Journal of Travel Research*, 40: 124–131.

McCrone, D., Morris, A. and Kelly, R. (1995) *Scotland – The Brand: The Making of Scottish Heritage*, Edinburgh: Edinburgh University Press.

McField, M., Wells, S. and Gibson, J. (eds) (1996) *State of the Coastal Zone Report Belize, 1995*, Belize City: Coastal Zone Management Programme/Government of Belize.

McGregor, J. (2002) 'The politics of disruption: war veterans and the local state in Zimbabwe', *African Affairs*, 101: 9–37.

MacIntyre, A. (1967) *A Short History of Ethics*, London: Routledge.

—— (1981) 'A crisis in moral philosophy', in D. Callaghan and H.T. Engelhardt Jr (eds) *The Roots of Ethics: Science, Religion, Values*, New York: Plenum Press.

—— (1993) *After Virtue: A Study in Moral Theory*, 2nd edn, London: Duckworth.

MacIntyre, N., Jenkins, J. and Booth, K. (2001) 'Global influences on access: the changing face of land access to public conservation lands in New Zealand', *Journal of Sustainable Tourism*, 9: 434–450.

MacKenzie, J.M. (1988) *The Empire of Nature: Hunting, Conservation and British Imperialism*, Manchester: Manchester University Press.

McKercher, B. (1993) 'Some fundamental truths about tourism: understanding tourism's social and environmental impacts', *Journal of Sustainable Tourism*, 1: 6–16.

Mackie, J. L. (1978) *Ethics: Inventing Right and Wrong*, Harmondsworth, UK: Penguin.

McMinn, S. and Cater, E. (1998) 'Tourist typology: observations from Belize', *Annals of Tourism Research*, 25: 675–699.

Macnaghten, P. and Urry, J. (1998) *Contested Natures*, London: Sage.

MacPherson, C.B. (1979) *The Political Theory of Possessive Individualism: Hobbes to Locke*, Oxford: Oxford University Press.

Malloy, D.C. and Fennell, D.A. (1998) 'Ecotourism and ethics: moral development and organizational cultures', *Journal of Travel Research*, 36(4): 47–56.

Mann, M. (2000) *The Community Tourism Guide*, London: Earthscan.

Mansperger, M.C. (1995) 'Tourism and cultural change in small-scale societies', *Human Organizations*, 54: 87–94.

Marfurt, E. (1997) 'Tourism and the Third World: dream or nightmare?', in L. France (ed.) *The Earthscan Reader in Sustainable Tourism*, London: Earthscan.

Markwell, K. (1997) 'Dimensions of photography in a nature-based tour', *Annals of Tourism Research*, 24: 131–155.

Marshall, G. (1986) 'The workplace culture of a licensed restaurant', *Theory, Culture and Society*, 3: 33–48.

Martin, R. (1986) *Communal Areas Management Programme for Indigenous Resources (CAMPFIRE) Revised Version*, CAMPFIRE Working Document 1/86, Harare: Branch of Terrestrial Ecology, DNPWLM.

—— (1994a) 'Alternative approaches to sustainable use', paper presented at Conference on Conservation through Sustainable Use of Wildlife, University of Queensland, 8–11 February 1994.

—— (1994b) *The Influence of Governance on Conservation and Wildlife Utilization: Alternative Approaches to Sustainable Use: What Does and Doesn't Work*, Harare: DNPWLM.

Marx, K. (1974) 'Critique of the Gotha Programme', in D. Fernbach (ed.) *Karl Marx: The First International and After: Political Writings*, vol. 3, Harmondsworth, UK: Penguin.

—— (1990) *Capital: A Critique of Political Economy*, vol. 1, London: Penguin.

Maslow, A. (1954) *Motivation and Personality*, New York: Harper and Row.

Mathieson, A. and Wall, G. (1982) *Tourism, Economic, Physical and Social Impacts*, Harlow, UK: Longman.

Matthews, H.G. and Richter, L.K. (1991) 'Political science and tourism', *Annals of Tourism Research*, 18: 120–135.

Meadows, D.H. (1972) *The Limits to Growth*, London: Pan.

Metcalfe, S. (1992a) *Community Natural Resource Management: How Non-governmental Organizations Can Support Co-management Conservation and Development Strategies between Government and the Public*, Harare: Zimbabwe Trust.

—— (1992b) *Planning for Wildlife in an African Savanna: A Strategy Based on the Zimbabwean Experience: Emphasizing Communities and Parks*, Harare: Zimbabwe Trust.

—— (1992c) *The CAMPFIRE Programme in Zimbabwe: 'Empowerment' versus 'Participation' in Natural Resource Management in the Masoka Community*, Harare: Zimbabwe Trust.

Meyer, C.A. (1992) 'A step back as donors shift institution building from the public to the "private" sector', *World Development*, 20: 1115–1126.

Mill, J.S. (1987a) 'Bentham', in A. Ryan (ed.) *Utilitarianism and Other Essays: J.S. Mill and Jeremy Bentham*, Harmondsworth, UK: Penguin.

—— (1987b) 'Utilitarianism', in A. Ryan (ed.) *Utilitarianism and Other Essays: J.S. Mill and Jeremy Bentham*, Harmondsworth, UK: Penguin.

Ministry of Environment and Tourism (1992) *Policy for Wildlife in Zimbabwe*, Harare: Ministry of Environment and Tourism and DNPWLM.

Ministry of Natural Resources (1990) *National Conservation Strategy*, Harare: Ministry of Natural Resources.

Ministry of Tourism and the Environment/Inter-American Development Bank (1998) *Tourism Strategy Plan for Belize*, Help for Progress/Blackstone Corporation: Belmopan, Belize/Toronto, Canada.

Mitchell, R.E. and Reid, D.G. (2001) 'Community integration: island tourism in Peru', *Annals of Tourism Research*, 28: 113–139.

Moreno, J. and Littrel, M.A. (2001) 'Negotiating tradition: tourism retailers in Guatemala', *Annals of Tourism Research*, 28: 658–685.

Moscardo, G., Morrison, A.M. and Pearce, P.L. (1996) 'Specialist accommodation and ecologically sustainable tourism', *Journal of Sustainable Tourism*, 4: 29–52.

Mowforth, R. and Munt, I. (1998) *Tourism and Sustainability: Dilemmas in Third World Tourism*, London: Routledge.

Mulhall, S. and Swift, A. (1993) *Liberals and Communitarians*, Oxford: Blackwell.

Munt, I. (1994a) 'The "other" post-modern tourism: culture, travel and the new middle classes', *Theory, Culture and Society*, 11: 101–123.

—— (1994b) 'Ecotourism or egotourism?', *Race and Class*, 36: 49–60.

Murombedzi, J. (1990) *The Need for Appropriate Local Level Common Property Resource Management Institutions in Communal Tenure Regimes*, Harare: CASS, University of Zimbabwe.

—— (1992) *Decentralization or Recentralization? Implementing CAMPFIRE in the Omay Communal Lands of Nyaminyami District*, CASS Natural Resource Management Working Paper 2, Harare: CASS, University of Zimbabwe.

Murphree, M. (1991) *Communities as Institutions for Resource Management*, CASS Occasional Paper Series, Harare: CASS, University of Zimbabwe.

—— (1992) *Ivory Production and Sales in Zimbabwe*, Harare: Branch of Terrestrial Ecology, DNPWLM.

—— (1995) 'Optimal principles and pragmatic strategies: creating an enabling politico-legal environment for community based natural resource management (CBNRM)', keynote address to the Conference of the Natural Resources Management Programme, SADC Technical Coordination Unit, Malawi, USAID–NRMP Regional Programme, Chobe, Botswana, 3 April 1995.

Murphree, M. and Cumming, D.H.M. (1991) 'Savanna land use: policy and practice in Zimbabwe', paper presented at the IUBS UNESCO/UNEP Conference Workshop, Economic Driving Forces and Constraints on Savannah Land Use, Nairobi, Kenya, 1991.

Murphy, L. (2001) 'Exploring social interactions of backpackers', *Annals of Tourism Research*, 28: 50–67.

Musil, R. (1995) *The Man without Qualities*, London: Picador.

Naess, A. (1972) 'The shallow and the deep ecology movement', *Inquiry*, 16: 95–100.

Nash, R. (1982) *Wilderness and the American Mind*, New Haven, CT: Yale University Press.

—— (1989) *The Rights of Nature: A History of Environmental Ethics*, Madison: University of Wisconsin Press.

Neale, G. (1999) *The Green Travel Guide*, London: Earthscan.

Nepal, S. (2000) 'Tourism in protected areas: the Nepalese Himalaya', *Annals of Tourism Research*, 27: 661–681.

Neumann, R.P. (1996) 'Dukes, earls and ersatz Edens: aristocratic nature preservationists in colonial Africa', *Environment and Planning D: Society and Space*, 14: 79–98.

—— (1997) 'Primitive ideas: protected area buffer zones and the politics of land in Africa', *Development and Change*, 28: 559–582.

Nietschmann, B. (1997) 'Protecting indigenous coral reefs and sea territories, Miskito Coast, RAAN, Nicaragua', in S. Stevens (ed.) *Conservation through Cultural Survival: Indigenous Peoples and Protected Areas*, Washington, DC: Island Press.

Niven, C., Noble, J., Forsyth, S. and Wheeler, T. (1999) *Sri Lanka*, London: Lonely Planet.

Noddings, N. (1984) *Caring*, Berkeley, CA: University of California Press.

Norman, R. (1983) *The Moral Philosophers: An Introduction to Ethics*, Oxford: Clarendon Press.

Norton, A. (1996) 'Experiencing nature: the reproduction of environmental discourse through safari tourism in East Africa', *Geoforum*, 27: 355–373.

Norton, G. (1994) 'The vulnerable voyager: new threats for tourism', *World Today*, 50: 237–239.

Oelschlaeger, M. (1991) *The Idea of Wilderness*, New Haven, CT: Yale University Press.

Opperman, M. (1993) 'Tourism space in developing countries', *Annals of Tourism Research*, 20: 535–556.

O'Riordan, T. and Jordan, A. (1995) 'The precautionary principle in contemporary environmental politics', *Environmental Values*, 4: 191–212.

O'Shaughnessy, H. (2000) *The Politics of Torture*, London: Latin American Bureau.

Page, S.J. and Dowling, R.K. (2002) *Ecotourism*, Harlow, UK: Prentice Hall.

Parekh, B. (1973) *Bentham's Political Thought*, London: Croom Helm.

Pattullo, P. (1996) *Last Resorts: The Cost of Tourism in the Caribbean*, London: Cassell.

Pearce, D., Markandaya, A. and Barbier, E.B. (1989) *Blueprint for a Green Economy*, London: Earthscan.

—— (1995) *Capturing Environmental Value*, London: Earthscan.

Pearce, P. and Moscardo, G. (1986) 'The concept of authenticity in touristic experience', *Australian and New Zealand Journal of Sociology*, 22: 122–132.

Pearce, P.L. (1995) 'From culture shock and culture arrogance to culture exchange: ideas towards sustainable socio-cultural tourism', *Journal of Sustainable Tourism*, 3: 143–154.

Peardon, T.P. (1974) 'Bentham's ideal republic', in B. Parekh (ed.) *Jeremy Bentham: Ten Critical Essays*, London: Frank Cass.

Peet, R. (1999) 'Social theory, post-modernism and the critique of development', in G. Benko and U. Strohmeyer (eds) *Space and Social Theory: Interpreting Modernity and Postmodernity*, Oxford: Blackwell.

Peperzak, A. (1992) *To the Other: An Introduction to the Philosophy of Emmanuel Levinas*, West Lafayette, IN: Purdue University Press.

Philip, J. and Mercer, D. (1999) 'Commodification of Buddhism in contemporary Burma', *Annals of Tourism Research*, 26: 21–54.

Picard, M. (1995) 'Cultural heritage and tourist capital: cultural tourism in Bali', in M.F. Lanfant, J.B. Allcock and E.M. Bruner (eds) *International Tourism: Identity and Change*, London: Sage.

Pieterse, J.N. (2001) *Development Theory: Deconstructions/Reconstructions*, London: Sage.

Pilger, J. (1992) *Distant Voices*, London: Vintage.

—— (1998) *Hidden Agendas*, London: Vintage.

Pi-Sunyer, O. (1989) 'Changing perceptions in a Catalan resort town', in V. Smith (ed.) *Hosts and Guests: The Anthropology of Tourism*, Philadelphia: University of Pennsylvania Press.

Pizam, A. and Mansfeld, Y. (eds) (1996) *Tourism, Crime and International Security Issues*, Chichester, UK: John Wiley.

Pizam, A. and Sussman, S. (1995) 'Does nationality affect tourist behaviour?', *Annals of Tourism Research*, 22: 901–917.

Place, S.E. (1991) 'Nature tourism and rural development in Tortuguero', *Annals of Tourism Research*, 18: 186–201.

Plato (1987) *The Republic*, Harmondsworth, UK: Penguin.

Plumwood, V. (1993) *Feminism and the Mastery of Nature*, London: Routledge.

—— (2002) *Environmental Culture: The Ecological Crisis of Reason*, London: Routledge.

Poirier, R.A. (1997) 'Political risk analysis and tourism', *Annals of Tourism Research*, 24: 675–686.

Porter, M. (1990) *The Competitive Advantage of Nations*, London: Macmillan.

Prentice, R. and Andersen, V. (2000) 'Evoking Ireland: modelling tourism propensity', *Annals of Tourism Research*, 27: 490–516.

Price, M.F. (ed.) (1996) *People and Tourism in Fragile Environments*, Chichester, UK: John Wiley.

Primarck, R.B., Bray, D., Galleti, H.A. and Ponciano, I. (eds) (1988) *Timber, Tourists and Temples: Conservation and Development in the Maya Forest of Belize, Guatemala and Mexico*, Washington, DC: Island Press.

Prosser, R. (1994) 'Societal change and the growth of alternative tourism', in E. Cater and G. Lowman (eds) *Ecotourism: A Sustainable Option*, Chichester, UK: John Wiley.

Raftopoulos, B. (2002) 'Briefing: Zimbabwe's 2002 presidential elections', *African Affairs*, 101: 413–426.

Ranger, T. (1999) *Voices from the Rocks: Nature, Culture and History in the Matopos Hills of Zimbabwe*, Oxford: James Currey.

Ratnapala, N. (1999) *Tourism in Sri Lanka: The Social Impact*, Ratmalana: Sarvodaya Vishva Lekha.

Rawls, J. (1973) *A Theory of Justice*, Oxford: Oxford University Press.

—— (1992) 'Justice as fairness: political not metaphysical', in S. Avineri and A. de-Shalit (eds) *Communitarianism and Individualism*, Oxford: Oxford University Press.

—— (2001) *Justice as Fairness: A Restatement*, Cambridge, MA: Harvard University Press.

Redclift, M. (1992) 'Sustainable development and global environmental change', *Global Environmental Change*, 2: 32–42.

Redfoot, D. (1984) 'Touristic authenticity, touristic angst, and modern reality', *Qualitative Sociology*, 7: 291–309.

Reef and Rainforest World Wide Adventure Travel: Belize. Web site: <http://www.reefrainfrst.com/belize.html> (accessed 19 March 1999).

Regan, T. (1988) *The Case for Animal Rights*, London: Routledge.

Rich, B. (1993) *Mortgaging the Earth: The World Bank, Impoverishment and the Crisis of Development*, London: Earthscan.

Richter, L.K. (1992) 'Political instability and tourism in the Third World', in D. Harrison (ed.) *Tourism and the Less Developed Countries*, London: Belhaven Press.

Robb, J.G. (1998) 'Tourism and legends: archaeology of heritage', *Annals of Tourism Research*, 25: 579–596.

Rojek, C. (1995a) 'Veblen, leisure and human need', *Leisure Studies*, 14: 73–86.

—— (1995b) *Decentering Leisure: Rethinking Leisure Theory*, London: Sage.

Rojek, C. and Urry, J. (eds) (1997) *Touring Cultures: Transformations of Travel and Theory*, London: Routledge.

Rorty, R. (1999) 'Human rights, rationality and sentimentality', in O. Savić (ed.) *The Politics of Human Rights*, London: Verso.

Rosenthal, J.H. (1999) *Ethics and International Affairs: A Reader*, 2nd edn, Washington, DC: Georgetown University Press.

Rostow, W. (1991) *The Stages of Economic Growth*, Cambridge: Cambridge Press.

Ruddick, S. (1989) *Maternal Thinking*, Boston: Beacon Press.

Ruskin, J. (1888) *The Stones of Venice: Introductory Chapters and Local Indices for the Use of Travellers While Staying in Venice and Verona*, Orpington: George Allen.

—— (2000) 'Landscape, mimesis and morality', an excerpt from 'Modern Painters' in L. Coupe (ed.) *The Green Studies Reader: From Romanticism to Ecocriticism*, London: Routledge.

Russell, B. (1930) *The Conquest of Happiness*, London: George Allen and Unwin.
Ryan, A. (ed.) (1987) *Utilitarianism and Other Essays: J.S. Mill and Jeremy Bentham*, Harmondsworth, UK: Penguin.
Ryan, C., Hughes, K. and Chirgwin, S. (2000) 'The gaze, the spectacle and ecotourism', *Annals of Tourism Research*, 27: 148–163.
Sagoff, M. (1989) *The Economy of the Earth: Philosophy, Law, and the Environment*, Cambridge: Cambridge University Press.
Said, E. (1978) *Orientalism*, London: Routledge.
San Pedro Sun (1997a) 'Punta Gorda – will Toledo learn what is right and what is wrong in conservation and tourism?', 3 May.
—— (1997b) 'Where have all the sardines gone?', 30 May.
—— (1997c) 'Tourist arrivals increase second quarter', 8 August.
—— (1997d) 'Ecotourism to become the fastest growing segment in world tourism', 31 October.
—— (1997e) 'Minister says Mexican air routes through Belize are not a done deal', 28 November.
Sandel, M. (1992) 'The procedural republic and the unencumbered self', in S. Avineri and A. de-Shalit (eds) *Communitarianism and Individualism*, Oxford: Oxford University Press.
Saqui, Pio (2000) Pers. comm. from Pio Saqui, Director of the Marine Research Centre, UCB, Belize City, 18 May.
Sarre, P. (1995) 'Towards global environmental values: lessons from western and eastern experience', *Global Environmental Values*, 4: 115–127.
Sasseen, J. (1993) 'Companies clean up', *International Management*, October, 30–31.
Scheyvens, R. (2002) 'Backpacker tourism and third world development', *Annals of Tourism Research*, 29: 144–164.
Schivelbusch, W. (1986) *The Railway Journey: The Industrialization of Time and Space in the Nineteenth Century*, Leamington Spa, UK: Berg.
Scholte, J. (2000) *Globalisation: A Critical Introduction*, London: Palgrave.
Schrader-Frechette, K. (1999) 'Chernobyl, global environmental injustice and mutagenic threats', in N. Low (ed.) *Global Ethics and the Environment*, London: Routledge.
Sellman, D. (1997) 'The virtues in the moral education of nurses: Florence Nightingale revisited', *Nursing Ethics*, 4(1): 3–11.
Selwyn, T. (1992) 'Tourism society and development', *Community Development Journal*, 27: 353–360.
—— (ed.) (1996) *The Tourist Image: Myths and Myth Making in Tourism*, Chichester: John Wiley.
Sen, A. (1977) 'Rational fools: a critique of the behavioural foundations of economic theory', *Philosophy and Public Affairs*, 6: 317–344.
—— (1999) *Development as Freedom*, Oxford: Oxford University Press.
Sen, A. and Williams, B. (eds) (1982) *Utilitarianism and Beyond*, Cambridge: Cambridge University Press.
Shaftesbury, Lord (1999) *Characteristics of Men, Manners, Opinions, Times*, Cambridge: Cambridge University Press.
Shephard, P. (1997) *The Others: How Animals Made Us Human*, Washington, DC: Island Press.
Shields, R. (ed.) *Lifestyle Shopping*, London: Routledge.
Shiva, V. (1989) *Staying Alive: Women, Ecology and Development*, London: Zed Books.
Shoman, A. (1995) *Backtalking Belize*, Belize City: Angelus Press.

Simmel, G. (1990) *The Philosophy of Money*, London: Routledge.

—— (1997) 'The Alpine journey', in D. Frisby and M. Featherstone (eds) *Simmel on Culture: Selected Writings*, London: Sage.

Simmons, M.S. (1999) 'Aboriginal heritage art and moral rights', *Annals of Tourism Research*, 27(2): 412–431.

Simpson, J. (1997) *Dark Shadows Falling*, London: Vintage.

Sinclair, M.T., Alizadeh, P. and Onunga, E.A.A. (1992) 'The structure of international tourism and tourism development in Kenya', in D. Harrison (ed.) *Tourism and the Less Developed Countries*, London: Belhaven Press.

Sindiga, I. (1999) *Tourism and African Development: Change and Challenge of Tourism in Kenya*, Aldershot, UK: Ashgate.

Singer, P. (1986) 'Animal liberation', in D. Van De Veer and C. Pierce (eds) *People, Penguins and Plastic Trees*, Belmont, CA: Wadsworth.

—— (1991) *Animal Liberation*, 2nd edn, London: HarperCollins.

Smart, J.J.C. and Williams, B. (1990) *Utilitarianism: For and Against*, Cambridge: Cambridge University Press.

Smith, D.M. (2000) *Moral Geographies: Ethics in a World of Difference*, Edinburgh: Edinburgh University Press.

Smith, J. (1994) 'Tourism and the environment: a *Zimbabwe Sun* perspective', *Southern African Economist*, Harare: SADC Press Trust.

Smith, M. (2000) 'Snap-shots: tourism and the environment', *Environmental Politics*, 9(4): 146–149.

—— (2001a) *An Ethics of Place: Radical Ecology, Postmodernity, and Social Theory*, Albany, NY: State University of New York Press.

—— (2001b) 'Environmental anamnesis: Walter Benjamin and the ethics of extinction', *Environmental Ethics*, 23(4): 359–376.

Smith, M.D. and Krannich, R.S. (1998) 'Tourism dependence and resident attitudes', *Annals of Tourism Research*, 25: 783–802.

Smith, V.L. (1989) *Hosts and Guests: The Anthropology of Tourism*, Philadelphia: University of Pennsylvania Press.

Sonmez, S.F. (1998) 'Tourism, terrorism and political instability', *Annals of Tourism Research*, 25: 416–456.

Sonmez, S.F. and Graefe, A.R. (1998) 'The influence of terrorism risk on foreign tourism decisions', *Annals of Tourism Research*, 25: 112–144.

Soulé, M.E. and Lease, G. (eds) (1995) *Reinventing Nature: Responses to Postmodern Deconstruction*, Washington, DC: Island Press.

Sport Diver Magazine (1997) 6, Wendens Ambo, Essex, UK: Market Link Publishing.

Spretnak, C. (1999) *The Resurgence of the Real: Body, Nature and Place in a Hypermodern World*, London: Routledge.

Stabler, M.J. (ed.) (1997) *Tourism and Sustainability: Principles to Practice*, Oxford: Oxford University Press.

Steele, P. (1995) 'Ecotourism: an economic analysis', *Journal of Sustainable Tourism*, 3: 29–44.

Stevens, S. (ed.) (1997) *Conservation through Cultural Survival: Indigenous Peoples and Protected Areas*, Washington, DC: Island Press.

Stiles, D. (1994) 'Tribals and trade: a strategy for cultural and ecological survival', *Ambio*, 23: 106–111.

Stoneman, C. and Cliffe, L. (1989) *Zimbabwe: Politics, Economics and Society*, London: Pinter.

Stonich, S. (1998) 'Political ecology of tourism', *Annals of Tourism Research*, 25: 25–54.

Stonich, S., Sorensen, J.H. and Hundt, A. (1995) 'Ethnicity, class and gender in tourism development: the case of the Bay Islands, Honduras', *Journal of Sustainable Tourism*, 3: 1–28.

Sunday Gazette (1995 'Communal sanctuary for rhino', 26 March.

Sunday Mail (South Africa) (1990) 'Ban on ivory to affect farmers', 8 July.

—— (1995) 'Meeting the challenge of the tourism industry', 26 March.

Swain, M.B. (1989) 'Gender roles in indigenous tourism: Kuna Mola, Kuna Yala, and cultural survival', in V.L. Smith (ed.) *Hosts and Guests: The Anthropology of Tourism*, Philadelphia: University of Pennsylvania Press.

Taylor, C. (1991) *The Ethics of Authenticity*, Cambridge, MA: Harvard University Press.

Taylor, J.P. (2001) 'Authenticity and sincerity in tourism', *Annals of Tourism Research*, 28: 7–26.

Teague, K. (2000) 'Tourist markets and Himalayan craftsmen', in M. Hitchcock and K. Teague (eds) *Souvenirs: The Material Culture of Tourism*, Aldershot, UK: Ashgate.

Teye, V., Sirakaya, E. and Sonmez, S.F. (2002) 'Residents' attitudes toward tourism development', *Annals of Tourism Research*, 29: 668–688.

The Times (UK) (2001) 'Brutal Mugabe rebuffs EU on election rights', 24 November.

Thompson, E.P. (1967) 'Time, work-discipline and industrial capitalism', *Past and Present*, 38: 56–97.

Toledo Maya Cultural Council and Toledo Alcaldes Association (1997) *Maya Atlas: The Struggle to Preserve Maya Land in Southern Belize*, Berkeley, CA: North Atlantic Books.

Tosun, C. (2002) 'Host perceptions of impacts: a comparative tourism study', *Annals of Tourism Research*, 29: 231–253.

Tourism Concern (1995) 'Forced labour and relocations in Burma', *Tourism in Focus*, 15: 6–7.

Tremlett, G. (2002) 'Benidorm gets high and mighty ugly', *Guardian*, 18 May.

Tuck, R. (1977) *Natural Rights Theories: Their Origin and Development*, Cambridge: Cambridge University Press.

Turnbull, C. (1973) *The Mountain People*, London: Jonathan Cape.

Urry, J. (1990) *The Tourist Gaze: Leisure and Travel in Contemporary Societies*, London: Sage.

—— (1991) 'The sociology of tourism', in C.P. Cooper (ed.) *Progress in Tourism, Recreation and Hospitality Management*, vol. 3, London: Belhaven Press.

—— (1992) 'The tourist gaze and the environment', *Theory, Culture and Society*, 9: 3–24.

—— (1994) 'Cultural change and contemporary tourism', *Leisure Studies*, 13: 233–238.

—— (1997) *The Tourist Gaze: Leisure and Travel in Contemporary Societies*, London: Sage.

Van De Veer, D. and Pierce, C. (eds) *People, Penguins and Plastic Trees*, Belmont, CA: Wadsworth.

Van den Berghe, P. (1994) *The Quest for the Other: Ethnic Tourism in San Cristóbal, Mexico*, Seattle, WA: University of Washington Press.

—— (1995) 'Marketing Mayas: ethnic tourism promotion in Mexico', *Annals of Tourism Research*, 22: 568–588.

Van den Berghe, P. and Flores Ochoa, J. (2000) 'Tourism and nativistic ideology in Cuzco, Peru', *Annals of Tourism Research*, 27: 7–26.

Vaux, T. (2001) *The Selfish Altruist: Relief Work in Famine and War*, London: Earthscan.

Vogel, S. (1996) *Against Nature: The Concept of Nature in Critical Theory*, Albany, NY: State University of New York Press.

Wain, K. (1995) *The Value Crisis: An Introduction to Ethics*, Msida, Malta: University of Malta.

Waitt, G. (2000) 'Consuming heritage: perceived historical authenticity', *Annals of Tourism Research*, 27: 835–862.

Waldron, J. (1987) *Nonsense upon Stilts: Bentham, Burke and Marx on the Rights of Man*, London: Methuen.

Wallace, G.N. and Pierce, S.M. (1996) 'An evaluation of ecotourism in Amazonas, Brazil', *Annals of Tourism Research*, 23: 843–873.

Wallerstein, I. (1979) *The Capitalist World-Economy*, New York: Cambridge University Press.

Wang, N. (1999) 'Rethinking authenticity in tourism experience', *Annals of Tourism Research*, 26: 349–370.

—— (2000) *Tourism and Modernity: A Sociological Analysis*, Oxford: Pergamon.

Warren, S. (1999) 'Cultural contestations at Disneyland Paris', in D. Crouch (ed.) *Leisure/Tourism Geographies: Practices and Geographical Knowledge*, London: Routledge.

Watson, G.L. and Kopachevsky, J.P. (1998) 'Interpretations of tourism as commodity', in Y. Apostolopoulos, S. Leivadi and A. Yiannakis (eds) *The Sociology of Tourism: Theoretical and Empirical Investigations*, London: Routledge.

Wearing, B. and Wearing, S. (1992) 'Identity and the commodification of leisure', *Leisure Studies*, 11: 3–18.

Weaver, D. (1994) 'Ecotourism in the Caribbean Basin', in E. Cater and G. Lowman (eds) *Ecotourism: A Sustainable Option*, Chichester, UK: John Wiley.

Weaver, D. and Elliot, K. (1996) 'Spatial patterns and problems in contemporary Namibian tourism', *Geographical Journal*, 162: 205–217.

Weber, M. (1964) 'Bureaucracy', in H.H. Gerth and C. Wright Mills (eds) *From Max Weber: Essays in Sociology*, London: Routledge and Kegan Paul.

—— (2001) *The Protestant Ethic and the Spirit of Capitalism*. London: Routledge.

Weed, T.J. (1994) 'Central America's peace parks and regional conflict resolution', *International Environmental Affairs*, 6: 175–190.

Weinberg, B. (1991) *War on the Land: Ecology and Politics in Central America*, London: Zed Books.

Wheat, S. (1994) 'Taming tourism', *Geographical Magazine*, 66: 16–19.

—— (1998) 'Ethical tourism: tourism concern'. Online. Available at: <http://www.mcb.co.uk/services/conferen/jan98/eit/1_wheat.html> (accessed 9 September 2000).

—— (2001) 'Fashion statement', *Tourism in Focus*, 39: 5.

Wickens, E. (2002) 'The sacred and the profane: a tourist typology', *Annals of Tourism Research*, 29: 834–851.

Wight, P. (1994) 'Environmentally responsible marketing of tourism', in E. Cater and G. Lowman (eds) *Ecotourism: A Sustainable Option*, Chichester: John Wiley.

Wildlife Society of Zimbabwe (1993) *Zimbabwe's Great Elephant Debate: A Report on Discussions Held during the AGM of the Wildlife Society of Zimbabwe at Hwange National Park on August 13, 1993*, Harare: Wildlife Society.

Wilkinson, P. (1992) 'Tourism – the curse of the nineties? Belize – an experiment to integrate tourism and the environment', *Community Development Journal*, 27: 386–395.

Williams, B. (1982) *Morality: An Introduction to Ethics*, Cambridge: Cambridge University Press.

Williams, J. and Lawson, R. (2001) 'Community issues and resident opinions of tourism', *Annals of Tourism Research*, 28: 269–290.

Williams, S. (1993a) 'Tourism on their own terms: how the TEA blends with village ways', *Belize Review*: 9–20.

—— (1993b) 'Pushing the boundaries of tourism', *Belize Review*: 3–8.

Williamson, J. (1993) 'Democracy and Washington consensus', *World Development*, 21(8): 132–136.

Wittgenstein, L. (1981) *Philosophical Investigations*, Oxford: Blackwell.

Wolmer, W. (2002) 'Wilderness gained, wilderness lost: wildlife management and land occupations in Zimbabwe's south-east lowveld', paper presented to the African Studies Association of the UK biennial conference, University of Birmingham, 9–11 September 2002.

Wood, R.E. (2000) 'Caribbean cruise tourism: globalization at sea', *Annals of Tourism Research*, 27: 345–370.

Woodhouse, P. (1992) 'Environmental degradation and sustainability', in T. Allen and A. Thomas (eds) *Poverty and Development in the 1990s*, Milton Keynes: Open University Press.

World Bank (1994) *World Development Report*, Oxford: Oxford University Press.

World Tourism Organization (2001) *Tourism Highlights 2001*, Madrid: WTO.

World Tourism Organization, UNEP and IUCN (1992) *Guidelines: Development of National Parks and Protected Areas for Tourism*, Technical Report Series 13, Nairobi: WTO/UNEP.

Worrall, S. (2002) 'Sold down the river', *Guardian Weekend*, 9 November: 26–35.

Yale, P. (1997) 'Is this conservation?', *Tourism in Focus*, 23: 8–9.

Young, B. (1983) 'Touristization of a traditional Maltese fishing–farming village: a general model', *Tourism Management* 4(1): 35–41.

Young, H. (1994) 'Eco-cultural tourism in Belize: reuniting man with the natural world', *Belize Review*: 4–6.

Zimbabwe Tourism Development Corporation (1993) *Proposed Marketing Policy and Strategy*, Harare: ZTDC.

Zimbabwe Trust (1992) *Wildlife: Relic of the Past or Resource of the Future? The Realities of Zimbabwe's Wildlife Policymaking and Management*, Harare: Zimbabwe Trust.

Zimbabwe Trust, DNPWLM and the CAMPFIRE Association (1990) *People, Wildlife and Natural Resources: The CAMPFIRE Approach to Rural Development*, Harare: WWF.

Zimmerman, M.E. (2000) 'The end of authentic selfhood in the post-modern age?', in M. Wrathall and J. Malpas (eds) *Heidegger, Authenticity, and Modernity: Essays in Honour of Hubert L. Dreyfus*, vol. 1, Cambridge, MA: MIT Press.

Index

access to land 82–3
aesthetics: globalization 18, 149; medieval 16; moral learning 48–9; values 9, 12, 157
Africa: animals *see* wildlife; safari industry 122, 146, 147–9; sub-Sahara 93; Zimbabwe *see* Zimbabwe
Aitchison, C. 112
alienation 42
Alps 48, 51, 52
alternative tourism 135
altruism 19
Ancient Greece 20, 38–9, 41, 42, 43–4
animal mistreatment 56, 64, 154–5
animals *see* wildlife
anti-foundationalism 110
Aristotle 44–5, 49, 92, 95
Aung San Suu Kyi, Daw 86
Australia 91, 126–7
authenticity: authentic holidays 59; culture 123–4; destination image 125–30; ethics 114–34, 165; heritage industry 114; meaning 114; nature 130–2, 147; North/South 115, 131; performed/staged authenticity 114–15, 116; politics 125–30

backpackers 50, 54, 124
Bali 126, 128
banana republics 131
Baudrillard, Jean 110, 111, 113, 163, 164
Bauman, Z. 2, 43, 47, 115, 165
Belize: agriculture 142–3; Blackstone Report 140; codes of conduct/ethics 88; community conservation 139–44; community-based ecotourism 139–40, 143; Coral Cay Conservation 50–1; coral reefs 120, 122, 123, 127, 131; craft production 128–9; diving 122–3, 124, 131; Eco-Tourism Association 88; economic diversification 138; El Pilar Archaeological Reserve 139; ethnic groups 144; hotel developments 65; illegal migrants 95; image and desire in tourism 120–5; Kekchi Council 129, 142; land claims 143–4; Maya 120, 123, 127, 128–30, 134, 139–44; Placenia 136; rainforest 120, 121, 123, 131, 142; Third World 3; Toledo Ecotourism Association (TEA) 140–2, 143, 144; Toledo Maya Cultural Council 143; Tour Guide Association 143; tour guides 41, 142–3, 144; Tourism Board (BTB) 121, 123, 131, 143; traditional dances 129; wildlife 139; wreck sites 122–3, 124
Benhabib, Seyla 104, 105, 106
Bentham, Jeremy 55, 56, 57, 58, 60, 61, 62, 64, 65, 66, 68, 69, 75–6, 77, 155
Botswana 149
Bourdieu, Pierre 62, 124
Brown, D. 117, 118
Buddhism 21, 35, 161
bullfighting 56
Burma 66, 86, 87
Burns, P. 85

Cal, Rafael 140, 141
Canada 64, 65
capitalism: business ethics 88; cultural logic 110; distributive justice 98; entrepreneurs 94; green capitalism 138; human rights 81; inequalities 94
Caribbean 95, 125
Carr, A.Z. 88–9
categorical imperatives 79–80, 89, 102
Central America 127–8, 130, 131, 143
charities 50

Chile 78
Ciulla, J.B. 88
codes of conduct/ethics: Belize 88;
    companies 88, 89; culturally embedded
    77; deontological guidelines 85; dress
    codes 35–6, 60; environment 78;
    mountaineering 52; WTO *see* World
    Tourism Organization
Cohen, J. 95
colonialism 83, 84, 91, 112, 125, 150
commodification: craft production 59, 129;
    culture 115; host-guest relationships
    16; labour 12; resistance 12, 30–1, 160;
    social life 16, 28, 112; traditional
    activities 59; wildlife 150, 157
communication and reciprocity 101–2
communicative ethics 99–106, 138
communitarianism 99, 101, 110
community-based tourism 56, 135,
    138–40, 143, 145–6
companies 78, 88, 89
conscience collective 32–3, 38
contingent valuation 26, 27
corporate responsibility 78, 88, 89
cost-benefit analysis 26, 27, 28
craft production 58, 59, 61, 128–9
culture: authenticity 123–4; codes of
    conduct/ethics 77; commodification
    115; cultural difference 32–4; cultural
    logic 110; cultural policy 59, 60;
    fragmentation 115; local exoticism 120

Dann, G. 117, 118–19, 120
decision-making 8, 60, 139
democratization 4
dependency theory 108
deregulation 88
developing countries: economics *see*
    economic development; good
    governance 4; under-development
    83–4; Western values 2, 83–4
differences: cultural difference 32–4;
    difference ethics 109–13; internal value
    differences 33, 38; other *see* otherness
Disney World 111
distributive justice 92, 97, 98, 138, 140,
    143, 144, 145, 164
division of labour 32, 33, 38
Doherty, F. 66
Donnelley, J. 81
Douzinas, C. 79
Doxey, G.V. 24
dress codes 35–6, 60
Drucker, Peter 89

Dryzek, J. 103–4
Duffy, R. 94, 95, 103, 139
Durkheim, Emile 15, 29, 30, 32, 33, 34,
    35, 36, 47, 53, 161

East Timor 78
economic development: modernization 4,
    83, 107; strategies 83–4; sustainable
    see sustainable development;
    utilitarianism 60; values 15
economics: diversification 4, 138; values
    8, 9, 13, 15, 16, 23–31
ecotourism: alternative tourism 135;
    community conservation 139–44;
    community-based ecotourism 139–40,
    143; eco-trails 140, 141; green
    capitalism 138; marketing 140
education: tourists 59, 60, 61; virtue 51,
    52, 54
ego-tourism 54, 116
egocentric monism 111
Egypt 119
El Salvador 127
El Tatio Geyser 78
Elsrud, T. 50, 54, 116, 124
emotional labour 40–1, 42
enclave tourism 138
Enlightenment 47, 76
environment: codes of conduct/ethics 78;
    deep ecology 154; degradation 136–7;
    responsibility 78; Rio Earth Summit
    (1992) 104, 137; wildlife conservation
    50–1, 55, 85, 139, 148, 150, 153
epistemic values 10
equal opportunities 97
equality 93, 94, 95, 96–7
Eritrea 85
Escobar, A. 83
ethical labour 45, 46, 47, 50, 51, 53–4
ethical tourism: sustainability 135–59, 163
ethics: ambiguities 2; authenticity 114–34,
    165; business ethics 88–9;
    communicative 99–106, 138; cultural
    difference 32–4; currency 161;
    difference ethics 109–13; discourses
    2–3, 103, 104, 105; ethics of care
    106–9, 110; evaluation 9; morals *see*
    morality; otherness 111–12, 124, 164,
    165, 166; sustainable tourism 135–59;
    values *see* values
Ethiopia: Sheraton Hotel 62, 63
ethnographic imagery 117
exchange-value 161–2
exotic 112, 114, 118, 119–20, 134

Fairweather, Mike 121, 131
fantasy ideals 117
feminism 106–9
Flores Ochoa, J. 105, 127, 139, 142
foreign exchange earnings 66, 137, 147
Foucault, Michel 45, 46
France: Declaration of the Rights of Man and the Citizen (1791) 73, 74, 75–6, 77, 81
freedom of movement 74

Gadamer, Hans-Georg 47, 166
The Gambia: sex tourism 41
gender relations: employment 15; moral judgements 107
genuine fakes 117
geography: morality 3
Gewirth, Alan 76
Gide, André 18
Gilligan, Carol 107, 108, 109, 110
globalization: aesthetics 18, 149; civil society 104; human rights 76–7; modernity 1–2, 5, 18; modernization policies 4
Godwin, William 64
Goethe, Johann Wolfgang von 47
golden mean 45
golf course developments 62–3, 73
governments: despotic regimes 78, 85; human rights 77–8, 79, 80, 83
Grand Tour 47–8, 49, 50, 54
Guatemala: Maya 127; religion 37; textiles 41, 95
Guyana 78

Habermas, Jürgen 76, 99, 101–6, 138, 139
Hadot, Paul 45–6
Hall, C.M. 92, 126, 128
happiness 53–72
hardships 54, 124
Harrison, D. 5, 128
Heckman, S. 107, 108
hedonism 55, 66, 116
hedonistic calculus 3, 57, 60, 61, 69, 155
Hegel, Georg Wilhelm Friedrich 47
heritage industry: authenticity 114; Scotland 2; slave trade 125; Wales 2
hierarchy of needs 24
Himba of Namibia 9–10
Hobbes, Thomas 18–19, 21
Homer 38–9, 43
Honduras 127
host-guest relationships:

commodification 16; common ethics 35; otherness 112; reciprocity 5
hotel developments 62, 63, 65, 135
human rights: abuse 78, 86, 87; basic liberties 97, 98; capitalism 81; communism 81; development aid 84; globalization 76–7; governments 77–8, 79, 80, 83; modernity 5

ideal speech situation 102, 103, 139
identity: morality 50, 54; regional 105, 126, 127; travel stories 116
independent travellers 50, 54, 61
individualism and autonomy 33, 53–4
Indonesia 40, 78, 126, 128
industrialization 137
International Association of Convention and Visitor Bureaus 74
International Monetary Fund (IMF): neoliberalism 4
International Tourism Exchange 141
Irigaray, Luce 110, 111, 112, 132, 166
irradex (index of irritation) 24
Islam 35

Jameson, Frederick 110
Jefferson, Thomas 73
Johnston, L. 108, 112
justice: distributive 92, 97, 98, 138, 140, 143, 145, 164; fairness 92, 93; non-distributive 92; social justice 91–9

Kant, Immanuel 74, 79–80, 81, 87, 89, 90, 95, 102, 109, 162, 165
Keefe, J. 62
Kenya 126, 146, 147–8, 149
Kohlberg, Lawrence 106–7
Krauss, Rosalind 54

labour: commodification 12; division of labour 32, 33, 38; emotional labour 40–1, 42; ethical labour 45, 46, 47, 50, 51, 53–4; volunteer labour 50–1; wage labour 13–14
landscape: access to land 82–3; imagery 118–19, 130–1, 149; mountains *see* mountaineering; parks *see* national parks
Lanfant, M.F. 115, 125, 126
Levinas, Emmanuel 110, 111, 112, 132, 134, 166
Lewis, Norman 10–11, 15
Locke, John 75, 76, 81, 95
Lonely Planet guides 86

Lukes, Steven 30, 67
Lyotard, Jean-François 110

Maasai 85, 126
MacCannell, D. 1, 15, 32, 35, 116
MacIntyre, Alasdair 18, 22, 29, 38, 42, 49, 76, 80, 82, 99, 164
Madagascar 119
Malawi 34
Malaysia 117
maldevelopment 84
Maoris 82–3, 126, 134
Marcus Aurelius 46
Markwell, K. 117–18
Marshall, Alfred 69
Marx, Karl Heinrich 42, 81, 94, 95, 163
Maslow, A. 24
mass tourism: cultural impact 135; morality 49–50; package holidays 61–2; utilitarianism 55–6, 66; wildlife 148
Maya 120, 123, 127, 128–30, 131, 134, 139–44
meta-ethical theories 55
Mexico 127
military repression 37–8
Mill, James 55
Mill, John Stuart 61, 62, 66, 69
modernity: globalization 1–2, 5, 18; human rights 5; late modernity 1, 115, 164; liquid modernity 165; tourism development 1–2; universality 110; utilitarianism 5; value fields 16, 17
modernization and economic development 4, 83, 107
money: currency 161; economic value 8–9, 16; exchange-value 161–2; personal preferences 25–6; utility 70
morality: amoralism 18–23, 24, 51, 52, 71; betterment 54; categorical imperatives 79–80, 89, 102; disagreement 38; diversity 34, 35, 36; identity 50, 54; mass tourism 49–50; moral outrage 11, 12, 13, 14, 19, 26, 27; moral reasoning 107; moral relativism 34–8, 44, 51, 52, 53, 81; moral scepticism 16–23, 38, 79, 116; moral subjectivism 23–31, 38, 51, 52; moral/ethical governance 54, 56, 65, 66; mountaineering 7, 52; norms 34; objectivism 29, 36, 44; publicness 90; social life 3, 13, 18, 22
mountaineering: access to land 82–3; alpinism 48, 51, 52; codes of conduct/ethics 52; morality 7, 52

Mundo Maya 127
Munt, I. 50, 54, 62, 116, 118, 149
Myanmar (Burma) 66, 86, 87

Naess, Arne 154
Namibia 9–10, 149
national parks: aesthetics *see* landscape; animals *see* wildlife; imagery 130–1, 149; Sri Lanka 58, 59; Zimbabwe 56, 94, 126, 148–9
natural environment and authenticity 130–2, 147
natural rights 75–6, 77, 87
neoclassical economics 25, 26
neoliberalism 4, 137, 144
New Zealand 82–3, 126, 134
Nightingale, Florence 40
non-governmental organizations (NGOs) 125, 153
norms 34
North/South: authenticity 115, 131; inequality 95; sustainable development 137, 138; values 3–4
Norway 82, 124
novelty 50

oppression 66, 67
organic solidarity 33, 38
original position 99, 101, 102, 103
otherness: binary oppositions 112; essentialization 114, 126; ethics 111–12, 124, 164, 165, 166; exotic 112, 114, 118, 119–20, 134; host-guest relationships 112; modernity 35; same 134; self-discovery 32

package holidays 50, 61–2, 66
Pareto optimality 70
participation in decision-making 60, 139
Pattullo, P. 5, 95, 125, 128, 130, 138
personal preferences 25–6, 38, 71
Peru 105, 160
Philippines: conservation projects 51; golf course developments 62–3; Ibaloi 62, 63, 64
photography 117–18, 146, 148, 149, 152, 155
Pigou, A.C. 69
Pilger, John 86
Plato 20, 44
Pollis, Adamantia 81
population displacement 62–3, 73, 86
post-modernism 2, 110–11, 115
poverty 84, 137

property development 62–3, 73, 86
public inquiries 103

qualities of pleasure 61, 62

Rabossi, Eduardo 76
railways 49, 50
Ratnapala, N. 57–60, 62
Rawls, John 91–2, 95–102, 110, 138, 164
Reagan, Ronald 88
regional identity 105, 126, 127
regional tourism 94, 149
religion: Buddhism 21, 35; conscience
    collective 32–3; Islam 35; Namibia
    9–10; sacrilege 37
resorts 135, 136, 138
Ridley, Jonathan 50
rights 73–85
risk 50, 54
Rorty, Richard 76
Rousseau, Jean Jacques 95
Ruskin, John 48, 49, 51, 54, 124
Russell, Bertrand 71

St Augustine of Hippo 46
Sandel, M. 99
Sasseen, J. 89
Schivelbusch, Wolfgang 49
Schmidt, Chet 141, 142, 144
Scotland 2
self-interest 18–19, 21
self-reflexivity 34–5, 36
semiotics 117
Sen, Amartya Kumar 69–70, 71
sex tourism: exploitation 92; The Gambia
    41; inequalities 96; South-East Asia 92;
    values 29
sexuality 15, 34
Shaftesbury, Anthony Ashley Cooper (3rd
    Earl) 46, 47
Sherwood, James B. 86
Shiva, Vandana 84, 161
Simmel, Georges 51, 52, 124
Simmons, M.S. 91
Simpson, Joe 7, 52
Singer, Peter 145, 155
slave trade 125
Smith, David M. 3, 24–5, 109, 110, 111,
    112, 125, 132
Smith, M. 107
social contract 75, 95, 110
social justice 91–9
social life and morality 3, 13, 18, 22
social life commodified 16, 28, 112

social roles: virtues 38–43, 53
social solidarity 15, 29, 32, 33
Sophists 44
South Africa 149
souvenirs 61, 129
Spanish Mediterranean coast 10–14, 135
sponsored activities 50
Sri Lanka: Minneriya National Park 58,
    59; sexuality 34; tourism development
    57–60
stoicism 46
sublime experiences 48
sustainable development:
    community-managed tourism 138–9;
    eco-tourism *see* ecotourism; legitimacy
    130; tourism 136–9; World Summit
    (2002) 137
sustainable tourism 135–59
Sweden 82

Tanzania 146, 149
Taylor, Charles 99
Taylor, J.P 126, 134
Thailand 123–4
Thatcher, Margaret Hilda 88
Third World 3, 81, 83, 84, 137
time management patterns transposed 1
tour guides: Belize 41, 142–3, 144;
    distributive justice 143, 144; licensing
    143, 144; stereotypical conversations
    41, 103
Tourism Concern 86, 104
tourism development: ambiguities 2, 5;
    benefits 142; interest groups 139;
    modernity 1–2; rights 81–5
tourism professionals 40
touristization 11
traditional activities 58, 59
travel brochures 118–20, 127, 128, 131
travel stories 116, 124
Tropical Paradise 112

United Kingdom 78
United Nations: Development Programme
    (UNDP) 70; Rio Earth Summit (1992)
    104, 137; Universal Declaration of
    Human Rights (1948) 74, 76
United States: business ethics 88;
    Declaration of Independence (1776) 73,
    74; democratic institutions 98; human
    rights abuse 78
Urry, J. 66, 111, 116, 117, 126, 132, 162,
    163, 164
utilitarianism: act and rule 66–7;

consequences 63–4, 66, 68; further
problems 68–9; hedonism 55, 66;
hedonistic calculus 3, 57, 60, 61, 69,
155; higher/lower pleasures 61;
impartiality 57; law 56; modernity 5;
quantity/quality 61; rationality 57;
social development 69; social
egalitarianism 62; theory 55–7, 161;
universality 57; utility 55, 56, 57–66,
70, 75; versatility 57; wildlife 55, 145

values: aesthetic *see* aesthetics; autonomy
13, 16; conflict 14–15; economics 8, 9,
13, 15, 16, 23–31; equal validity 34–5;
ethics 7–31; exchange-value 161–2;
inter-relations 13, 14; internal value
differences 33, 38; irrationality 27–8;
North/South 3–4; value fields 16, 17;
value spheres 10, 15, 18
Van den Berghe, P. 105, 127, 139, 142
Vanuatu 100–1
vegetarianism 36–7
Venice 48
virtues: elitism 49; humanity 43–52, 92;
professions 40; social roles 38–43, 53;
virtuous traveller 32–52
Vogel, Steven 106
volunteer work 50–1

wage labour 13–14
Wain, K. 80
Wales 2
Washington consensus 4
waste disposal 135, 136
Weber, Max 14, 77
Western values 2, 83–4
whale watching 64, 65
Wilde, Oscar 8, 9, 28
wildlife: animal rights 145, 154, 155;
animal welfare lobby 155;
anthropocentricity 150, 154; Belize
139; biodiversity 155–6;
commodification 150, 157;
community-based tourism 55, 145–6;
conservation 50–1, 55, 85, 139, 148,
150, 153; consumptive use 145, 146,
148, 150; Convention on the
International Trade in Endangered

Species (CITES) 156; ethical
treatment 145; mass tourism 148;
non-consumptive use 146, 152; parks
see national parks; photographic
tourism 146, 148, 149, 152, 155;
preservation value 156; safari
industry 122, 146, 147–9; sport
hunting 55, 136, 145, 146, 148, 149,
150–8; sustainable utilization 145–6,
147–8, 153–4, 157–8; utilitarianism
55, 145; viewing value 147–8, 152;
whale watching 64, 65; zero-use option
157
Williams, Bernard 36
willing to accept (WTA) 26, 27
willingness to pay (WTP) 26, 27
workaholics 42
World Bank 4, 84
World Health Organization (WHO) 93–4
World Tourism Organization: Global Code
of Ethics for Tourism 74, 77, 85, 87;
statutes 85; sustainable development
138; Tourism Bill of Rights and Tourist
Code 74, 77; World Committee on
Tourism Ethics 87

Young, Henry 131

Zimbabwe: adaptive management 151;
agriculture 138, 146–7; Association of
Tour and Safari Operators (ZATSO)
152, 156; CAMPFIRE programme 98,
136, 145–58; Communal Lands 146–7,
150, 151, 155; community-based
tourism 55, 145–6; Council for
Tourism 148; craft production 61;
economic diversification 138;
Gonarezhou National Park 148; ivory
trade 109, 156–7; Matopos National
Park 126; national park charges 94,
149; Parks Department 148, 149, 152,
153, 156–7; political disturbance 147,
149; sport hunting 55, 136, 145, 146,
148, 149, 150–8; Third World 3;
Tourism Development Corporation
(ZTDC) 147; trophy fees 152, 156;
ZANU-PF 147; Zimbabwe Sun Group
148

ESSENTIAL READING

Tourism and Sustainability, 2nd edition
Development and new tourism in the Third World
Martin Mowforth and Ian Munt

Hb: 0–415–27168–1
Pb: 0–415–27169–X   Routledge

Ecotourism
An introduction
David A. Fennell

Hb: 0–415–30364–8
Pb: 0–415–30365–6   Routledge

Issues in Cultural Tourism Studies
Melanie Smith

Hb: 0–415–25637–2
Pb: 0–415–25638–0   Routledge

Qualitative Research in Tourism
Jenny Phillimore and Lisa Goodson
Contemporary Geographies of Leisure, Tourism and Mobility series

Hb: 0–415–28086–9
Pb: 0–415–28087–7   Routledge

The Moralisation of Tourism
Sun, sand . . . and saving the world?
Jim Butcher
Contemporary Geographies of Leisure, Tourism and Mobility series

Hb: 0–415–29655–2
Pb: 0–415–29656–0   Routledge

Information and ordering details
For price availability and ordering visit our website **www.tandf.co.uk**
Subject Web address: **www.geographyarena.com**
Alternatively our books are available from all good bookshops.